LINEAR OPTIMIZATION PROBLEMS WITH INEXACT DATA

LINEAR OPTIMIZATION PROBLEMS WITH INEXACT DATA

M. FIEDLER
Academy of Sciences, Prague, Czech Republic

J. NEDOMA
Academy of Sciences, Prague, Czech Republic

J. RAMÍK
Silesian University, Karviná, Czech Republic

J. ROHN
Academy of Sciences, Prague, Czech Republic

K. ZIMMERMANN
Academy of Sciences, Prague, Czech Republic

 Springer

ISBN 978-1-4419-4094-0

e-ISBN: 0-387-32698-7

e-ISBN 978-0-387-32698-6

Printed on acid-free paper.

AMS Subject Classifications: 90C05, 90C60, 90C70, 15A06, 65G40

Printed in the United States of America.

9 8 7 6 5 4 3 2 1

springer.com

To our wives

Eva, Libuše, Eva, Helena and Olga

Contents

Preface

If we could make statistics of how computing time is distributed among individual mathematical problems, then, not considering database algorithms such as searching and sorting, the linear programming problem would probably be positioned on top. Linear programming is a natural and simple problem:

$$\text{minimize } c^T x \tag{0.1}$$

subject to

$$Ax = b, \tag{0.2}$$
$$x \geq 0. \tag{0.3}$$

In fact, we look for the minimum of a linear function $c^T x$, called the objective function, over the solution set of the system (0.2), (0.3), called the set of feasible solutions. As shown in linear programming textbooks, similar problems involving maximization or inequality constraints, or those missing (partly or entirely) the nonnegativity constraint, can be rearranged in the form (0.1)–(0.3) which we consider standard in the sequel.

It may seem surprising that such an elementary problem had not been formulated at the early stages of linear algebra in the 19th century. On the contrary, this is a typical problem of the 20th century, born of practical needs. As early as in 1902 J. Farkas [34] found a necessary and sufficient condition for solvability of the system (0.2), (0.3), called now the Farkas lemma. The linear programming problem attracted the interest of mathematicians during and just after World War II, when methods for solving large problems of linear programming were looked for in connection with the needs of logistic support of U.S. Armed Forces deployed overseas. It was also the time when the first computers were constructed.

An effective method for solving linear programming problems, the so-called simplex method, was invented in 1947 by G. Dantzig who also created a unified theory of linear programming [31]. In this context the name of the Soviet mathematician L. V. Kantorovich should be mentioned whose fundamental work had emerged as early as in 1939; however, it had not become known to the Western scientists till 1960 [66]. In the 'fifties, the methods of linear programming were applied enthusiastically as it was supposed that they could manage to create and resolve national economy plans. The achieved results, however, did not satisfy the expectations. This caused some disillusion and in the 'sixties a stagnation in development of mathematical methods and models had occurred which also led to the loss of belief in the power of computers. There were various reasons for the fact that the results of linear programming modeling did not often correspond to the expectations of the planners. One of them, which is the central topic of this book, was inexactness of the data, a phenomenon inherent in most practical problems.

Before we deal with this problem, let us finalize our historical excursion. The new wave of interest concerning linear programming emerged at the end of the 'seventies and at the beginning of the 'eighties. By that time, complexity of the linear programming problem was still unresolved. It was conjectured that it might be NP-hard in view of the result by V. Klee and G. Minty [72] who had shown by means of an example that the simplex method may take an exponential number of steps. In 1979, L. G. Khachian [71] disproved this conjecture by his ellipsoid method which can solve any linear programming problem in polynomial time. Khachian's result, however, was still merely of theoretical importance since in practical problems the simplex method behaved much better than the ellipsoid method. Later on, in 1984, N. Karmarkar [67] published his new polynomial-time algorithm for linear programming problems, a modification of a nonlinear programming method, which could substitute the simplex method. Whereas the simplex method begins with finding some vertex of the convex polyhedron and then proceeds to the neighboring vertices in such a way that the value of the objective function decreases up to the optimal value, Karmarkar's method finds an interior point of the polyhedron and then goes through the interior towards the optimal solution.

Optimization problems in finite-dimensional spaces may be characterized by a certain number of fixed input parameters that determine the structure of the problem in question. For instance, in linear programming problems such fixed input parameters are the coefficients of the objective function, of the constraint matrix and of the right-hand sides of the constraints. The solution of such optimization problems consists in finding an optimal solution for the given fixed input parameters. One of the reasons for the "crisis" of linear programming in the 'sixties and 'seventies was the uselessness of the computed solutions—results of the linear programming models—for practical decisions. The coefficients of linear programming models are often not known exactly, elicited by inexact methods or by expert evaluations, or, in other words, the nature of the coefficients is vague. For modeling purposes we usually use "average" values of the coefficients. Then we obtain an optimal solution of the model that is not always optimal for the original problem itself.

One of the approaches dealing with inexact coefficients in linear programming problems and trying to incorporate the influence of imprecise coefficients into the model is stochastic linear programming. The development of this area belongs to the 'sixties and 'seventies and is connected with the names of R. J. Wets, A. Prékopa and K. Kall.

The stochastic programming approach may have two practical disadvantages. The first one is associated with the numerics of the transformation of the stochastic linear programming problem to the deterministic problem of nonlinear programming. It is a well-known fact that nonlinear programming algorithms are practically applicable only to problems of relatively small dimensionality. The basic assumption of stochastic linear programming problems is that the probability distributions (i.e., distribution functions, or density functions) are known in advance. This requirement is usually not satisfied. The coefficients are imprecise and the supplementary information does not have a stochastic nature. More often, they are estimated by experts, eventually supplemented by the membership grades of inexactness or vagueness in question.

The problem of linear programming with inexact data is formulated in full generality as follows,

$$\text{minimize } c^T x \tag{0.4}$$

subject to

$$Ax = b, \, x \geq 0, \tag{0.5}$$

$$A \in \mathbf{A}, \, b \in \mathbf{b}, \, c \in \mathbf{c}, \tag{0.6}$$

where $\mathbf{A}$, $\mathbf{b}$ and $\mathbf{c}$ are subsets of $\mathbb{R}^{m \times n}$, $\mathbb{R}^m$ and $\mathbb{R}^n$, respectively, expressing inexactness of the data, and x is an n-dimensional vector.

Various approaches to this problem have been developed in the past with different descriptions of changes in the input data. In the frame of this book we deal with three approaches important from the practical point of view.

One of the basic research tools for investigation of linear optimization problems with inexact data (0.4)–(0.6) is matrix theory. In Chapter 1 basic notions on matrices and determinants, as well as on norms and basic numerical algebra are presented. Special attention is paid to such topics as symmetric matrices, generalized inverses, nonnegative matrices and M- and P-matrices.

In the literature, sufficient interest has not been devoted to linear programming problems with data given as intervals. The individual results, interesting by themselves, do not create a unified theory. This is the reason for summarizing existing results and presenting new ones within a unifying framework. In Chapter 2 solvability and feasibility of systems of interval linear equations and inequalities are investigated. Weak and strong solvability and feasibility of linear systems $Ax = b$ and $Ax \leq b$, where $A \in \mathbf{A}$ and $b \in \mathbf{b}$, are studied separately. In this way, combining weak and strong solvability or feasibility of the above systems we arrive at eight decision problems. It is shown that all of them can be solved by finite means, however, in half of the cases the number of steps is exponential in matrix size and the respective problems are proved to be NP-hard. The other four decision problems can be solved in polynomial time. The last part of the chapter is devoted to special types of solutions (tolerance, control and algebraic solutions), and to the square case.

Chapter 3 deals with the interval linear programming problem (0.4)–(0.6) where $\mathbf{A} = [\underline{A}, \overline{A}]$ is an interval matrix and $\mathbf{b} = [\underline{b}, \overline{b}]$, $\mathbf{c} = [\underline{c}, \overline{c}]$ are interval vectors. The main topics of the chapter are computation and properties of the exact lower and upper bounds of the range of the optimal value of the problem (0.4)–(0.6) with data varying independently of each other in the prescribed intervals. It is shown that computing the lower bound of the range can be performed in polynomial time, whereas computing the upper bound is NP-hard.

By generalizing linear programming problems with interval data, we obtain problems (0.4)–(0.6) with $\mathbf{A}$, $\mathbf{b}$ and $\mathbf{c}$ being compact convex sets. Such problems are studied in Chapter 4. In comparison with Chapter 3, $\mathbf{A}$, $\mathbf{b}$ and $\mathbf{c}$ are not necessarily matrix or vector intervals. Such a family of linear programming problems is called a linear programming problem with set coefficients (LPSC problem). Our interest is focused on the case where $\mathbf{A}$, $\mathbf{b}$ and $\mathbf{c}$ are either compact convex sets or, in particular, convex polytopes. We are interested primarily in systems of inequalities $Ax \leq b$ in (0.5) and later also in systems of equations. Under general assumptions, the usual form of the weak duality theorem is derived. Based on the previous results, the strong duality theorem is formulated and proved. The last part of Chapter 4 deals with algorithmic questions of LPSC problems: two algorithms for solving LPSC problems are proposed. Both algorithms are in fact generalizations of the simplex method.

A further generalization of linear programming problems with inexact data is a situation where coefficients of $\mathbf{A}$, $\mathbf{b}$ and $\mathbf{c}$ are associated with membership functions as a "degree of possibility", being a value from the unit interval $[0, 1]$. Then the sets $\mathbf{A}$, $\mathbf{b}$ and $\mathbf{c}$ are viewed as fuzzy subsets of the corresponding Euclidean vector spaces and the resulting linear programming problem is a linear programming problem with fuzzy coefficients. It is clear that the above linear programming problems with inexact coefficients are particular cases of problems with fuzzy coefficients. In Chapter 5 we propose a new general approach to fuzzy single- and multicriteria linear programming problems. A unifying concept of this approach is the concept of a fuzzy relation, particularly fuzzy extension of the usual inequality or equality relations. In fuzzy multicriteria linear programming problems the distinction between criteria and constraints can be modeled by various aggregation operators. The given goals can be achieved by the criteria whereas the constraints can be satisfied by the constraint functions. Both the feasible solution and compromise solution of such problems are fuzzy subsets of $\mathbb{R}^n$. On the other hand, the α-compromise solution is a crisp vector, as well as the max-compromise solution, which is, in fact, the α-compromise solution with the maximal membership degree. We show that the class of all multicriteria linear programming problems with crisp parameters can be naturally embedded into the class of fuzzy multicriteria linear programming problems with fuzzy parameters. It is also shown that the feasible and compromise solutions are convex under some mild assumptions and that a max-compromise solution can be found as the usual optimal solution of some classical multicriteria linear programming problem. The approach is demonstrated on a simple numerical example.

In the previous chapters, mostly linear systems of equations and inequalities as well as linear optimization problems with inexact interval data were investigated. The investigation took advantage of some well-known properties of linear systems and linear problems with exact data. Linear optimization problems are special convex optimization problems in which each local minimum is at the same time global. In Chapter 6, we investigate another class of optimization problems, the special structure of which makes it possible to find global optimal solutions. These problems form a special class of the so-called max-separable optimization problems. Functions occurring in these problems both as objective functions and in the constraints can be treated as "linear" with respect to a pair of semigroup operations. Properties of such optimization problems with interval data are presented as well.

In this research monograph we focus primarily on researchers as possible readers, or, the audience, particularly in the areas of operations research, optimization theory, linear algebra and eventually, fuzzy sets. However, the book may also be of some interest to advanced or postgraduate students in the respective areas.

Similar results to those published in this book, particularly concerning LP problems with interval uncertainty, have been published by Ben-Tal and Nemirovski et al.; see [14] to [20]. However, most of the results published in this monograph have been already published independently earlier in various journals and proceedings mainly between 1994 and 2000; see the list of references at the end of the book.

Chapter 1 was written by M. Fiedler, Chapters 2 and 3 by J. Rohn, Chapter 6 by K. Zimmermann, and Chapter 5 by J. Ramík who also wrote the part of Chapter 4 dedicated to the work of our colleague and friend Dr. Josef Nedoma who had started the work with us but was not able to conclude it, having passed away in July 2003.

The work on this monograph was supported during the years 2001 through 2003 by the Czech Republic Grant Agency under grant No. 201/01/0343.

Prague and Karviná   *Miroslav Fiedler, Jaroslav Ramík,*
                                             Jiří Rohn and Karel Zimmermann

1

Matrices

M. Fiedler

1.1 Basic notions on matrices, determinants

In this introductory chapter we recall some basic notions from matrix theory that are useful for understanding the more specialized sequel. We do not prove all assertions. The interested reader may find the omitted proofs in general matrix theory books, such as [35], [86], and others.

A *matrix of type m-by-n* or, equivalently, an $m \times n$ matrix, is a two-dimensional array of mn numbers (usually real or complex) arranged in m *rows* and n *columns* (m, n positive integers):

$$\begin{pmatrix} a_{11} & a_{12} & a_{13} & \cdots & a_{1n} \\ a_{21} & a_{22} & a_{23} & \cdots & a_{2n} \\ \cdot & \cdot & \cdot & \cdots & \cdot \\ a_{m1} & a_{m2} & a_{m3} & \cdots & a_{mn} \end{pmatrix} . \tag{1.1}$$

We call the number a_{ik} the *entry* of the matrix (1.1) in the ith row and the kth column. It is advantageous to denote the matrix (1.1) by a single symbol, say A, C, etc. The set of $m \times n$ matrices with real entries is denoted by $\mathbb{R}^{m \times n}$. In some cases, $m \times n$ matrices with complex entries will occur and their set is denoted analogously by $\mathbb{C}^{m \times n}$. In some cases, entries can be polynomials, variables, functions, etc.

In this terminology, matrices with only one column (thus, $n = 1$) are called *column vectors*, and matrices with only one row (thus, $m = 1$) *row vectors*. In such a case, we write $\mathbb{R}^m$ instead of $\mathbb{R}^{m \times 1}$ and –unless said otherwise– vectors are always column vectors.

Matrices of the same type can be added entrywise: if $A = (a_{ik})$, $B = (b_{ik})$, then $A + B$ is the matrix $(a_{ik} + b_{ik})$. We also admit multiplication of a matrix by a number (real, complex, a parameter, etc.). If $A = (a_{ik})$ and if α is a number (also called *scalar*), then αA is the matrix (αa_{ik}) of the same type as A.

An $m \times n$ matrix $A = (a_{ik})$ can be multiplied by an $n \times p$ matrix $B = (b_{k\ell})$ as follows: it is the $m \times p$ matrix $C = (c_{i\ell})$, where

$$c_{i\ell} = a_{i1}b_{1\ell} + a_{i2}b_{2\ell} + \cdots + a_{in}b_{n\ell}.$$

It is important to notice that the matrices A and B can be multiplied (in this order) only if the number of columns of A is the same as the number of rows in B. Also, the entries of A and B should be multiplicable. In general, the product AB is not equal to BA, even if the multiplication of both products is possible. On the other hand, the multiplication fulfills the *associative law*

$$(AB)C = A(BC)$$

as well as (in this case, two) *distributive laws*:

$$(A + B)C = AC + BC$$

and

$$A(B + C) = AB + AC,$$

whenever multiplications are possible.

Of basic importance are the *zero matrices*, all entries of which are zeros, and the *identity matrices*; these are *square* matrices, i.e., $m = n$, and have ones in the *main diagonal* and zeros elsewhere. Thus

$$(1), \quad \begin{pmatrix} 1 & 0 \\ 0 & 1 \end{pmatrix}, \quad \begin{pmatrix} 1 & 0 & 0 \\ 0 & 1 & 0 \\ 0 & 0 & 1 \end{pmatrix}$$

are identity matrices of *order* one, two and three. We denote zero matrices simply by 0, and the identity matrices by I, sometimes with a subscript denoting the order.

The identity matrices of appropriate orders have the property that

$$AI = A \text{ and } IA = A$$

hold for any matrix A.

Let now $A = (a_{ik})$ be an $m \times n$ matrix and let $\mathcal{M}$, $\mathcal{N}$, respectively, denote the sets $\{1, \ldots, m\}$, $\{1, \ldots, n\}$. If $\mathcal{M}_1$ is an *ordered* subset of $\mathcal{M}$, i.e., $\mathcal{M}_1 = \{i_1, \ldots, i_r\}$, $i_1 < \cdots < i_r$, and $\mathcal{N}_1 = \{k_1, \ldots, k_s\}$ an ordered subset of $\mathcal{N}$, then $A(\mathcal{M}_1, \mathcal{N}_1)$ denotes the $r \times s$ *submatrix* of A obtained from A by leaving the rows with indices in $\mathcal{M}_1$ and removing all the remaining rows and leaving the columns with indices in $\mathcal{N}_1$ and removing the remaining columns.

Particularly important are submatrices corresponding to consecutive row indices as well as consecutive column indices. Such a submatrix is called a *block* of the original matrix. We then obtain *partitioning* of the matrix A into blocks by splitting the set of row indices into subsets of the first, say, p_1 indices, then the set of the next p_2 indices, etc., up to the last p_u indices, and similarly splitting the set of column indices into subsets of consecutive $q_1, \ldots, q_v$ indices. If A_{rs} denotes the block describing the $p_r \times q_s$ submatrix of A obtained by this procedure, A can be written as

$$A = \begin{pmatrix} A_{11} & A_{12} & \cdots & A_{1v} \\ A_{21} & A_{22} & \cdots & A_{2v} \\ \cdot & \cdot & \cdots & \cdot \\ A_{u1} & A_{u2} & \cdots & A_{uv} \end{pmatrix}.$$

If, for instance, we partition the 3×4 matrix (a_{ik}) with $p_1 = 2$, $p_2 = 1$, $q_1 = 1$, $q_2 = 2$, $q_3 = 1$, we obtain the block matrix

$$\begin{pmatrix} A_{11} & A_{12} & A_{13} \\ A_{21} & A_{22} & A_{23} \end{pmatrix},$$

where, say A_{12} denotes the block $\begin{pmatrix} a_{12} & a_{13} \\ a_{22} & a_{23} \end{pmatrix}$.

On the other hand, we can form matrices from blocks. We only have to fulfill the condition that all matrices in each block row must have the same number of rows and all matrices in each block column must have the same number of columns.

The importance of block matrices lies in the fact that we can multiply block matrices in the same way as before:

Let $A = (A_{ik})$ and $B = (B_{k\ell})$ be block matrices, A with m block rows and n block columns, and B with n block rows and p block columns. If (and that is crucial) the first block column of A has the same number of columns as the first block row of B has the number of rows, the second block column of A has the same number of columns as the second block row of B has the number of rows, etc., till the number of columns in the last block column of A matches the number of rows in the last block row of B, then the product $C = AB$ is the matrix $C = (C_{i\ell})$, where

$$C_{i\ell} = A_{i1}B_{1\ell} + A_{i2}B_{2\ell} + \cdots + A_{in}B_{n\ell}.$$

Observe that the products $A_{ik}B_{k\ell}$ exist and can then be added.

Now let $A = (a_{ik})$ be an $m \times n$ matrix. The $n \times m$ matrix $C = (c_{pq})$ for which $c_{pq} = a_{qp}$, $p = 1, \ldots, n$, $q = 1, \ldots, m$, is called the *transpose matrix* to A. It is denoted by A^T. If A and B are matrices that can be multiplied, then

$$(AB)^T = B^T A^T.$$

Also,

$$(A^T)^T = A$$

for every matrix A.

This notation is also advantageous for vectors. We usually denote the column vector u with entries (coordinates) $u_1, \ldots, u_n$ as $(u_1, \ldots, u_n)^T$.

Of crucial importance are square matrices. If of fixed order, say n, and over a fixed field, e.g., $\mathbb{R}$ or $\mathbb{C}$, they form a set that is *closed* with respect to addition and multiplication as well as transposition. Here, closed means that the result of the operation again belongs to the set.

Some special types of square matrices are worth mentioning. A square matrix $A = (a_{ik})$ of order n is called *diagonal* if $a_{ik} = 0$ whenever $i \neq k$. Such a matrix is usually described by its diagonal entries as diag$\{a_{11}, \ldots, a_{nn}\}$. The matrix A is called *lower triangular* if $a_{ik} = 0$, whenever $i < k$, and *upper triangular* if $a_{ik} = 0$, whenever $i > k$. We have then:

Observation 1.1. *The set of diagonal (resp., lower triangular, resp., upper triangular) matrices of fixed order over a fixed field $\mathbb{R}$ or $\mathbb{C}$ is closed with respect to both addition and multiplication.*

A matrix A (necessarily square!) is called *nonsingular* if there exists a matrix C such that $AC = CA = I$. This matrix C (which can be shown to be unique) is called the *inverse matrix to* A and is denoted by A^{-1}. Clearly,

$$(A^{-1})^{-1} = A.$$

Observation 1.2. *If A, B are nonsingular matrices of the same order, then their product AB is also nonsingular and*

$$(AB)^{-1} = B^{-1}A^{-1}.$$

Observation 1.3. *If A is nonsingular, then A^T is nonsingular and*

$$(A^T)^{-1} = (A^{-1})^T.$$

Let us recall now the notion of the *determinant* of a square matrix $A = (a_{ik})$ of order n. We denote it as $\det A$:

$$\det A = \sum_{P=(k_1,\ldots,k_n)} \sigma(P) a_{1k_1} a_{2k_2} \cdots a_{nk_n},$$

where the sum is taken over all permutations $P = (k_1, k_2, \ldots, k_n)$ of the indices $1, 2, \ldots, n$, and $\sigma(P)$, the *sign* of the permutation P, is 1 or -1, according to whether the number of pairs (i, j) for which $i < j$ but $k_i > k_j$, is even or odd.

We list some important properties of the determinants.

Theorem 1.4. *Let $A = (a_{ik})$ be a lower triangular, upper triangular, or diagonal matrix of order n. Then*

$$\det A = a_{11} a_{22} \ldots a_{nn}.$$

In particular,

$$\det I = 1 \tag{1.2}$$

for every identity matrix.

We denote here, and in the sequel, the number of elements in a set S by card S. Let A be a square matrix of order n. Denote, as before, $\mathcal{N} = \{1, \ldots, n\}$. Whenever $\mathcal{M}_1 \subset \mathcal{N}$, $\mathcal{M}_2 \subset \mathcal{N}$, card $\mathcal{M}_1 =$ card $\mathcal{M}_2$, the submatrix $A(\mathcal{M}_1, \mathcal{M}_2)$

is square. We then call det $A(\mathcal{M}_1, \mathcal{M}_2)$ the *subdeterminant* or *minor* of the matrix A. If $\mathcal{M}_1 = \mathcal{M}_2$, we speak about *principal minors* of A.

Also, we speak about the *complementary submatrix* $A(\mathcal{N}\backslash\mathcal{M}_1, \mathcal{N}\backslash\mathcal{M}_2)$ of the submatrix $A(\mathcal{M}_1, \mathcal{M}_2)$ in A. For $\mathcal{M} \subset \mathcal{N}$, denote by $s(\mathcal{M})$ the sum of all numbers in $\mathcal{M}$. The determinant of the complementary submatrix multiplied by $(-1)^{s(\mathcal{M}_1)+s(\mathcal{M}_2)}$ is then called the *algebraic complement* of the subdeterminant det $A(\mathcal{M}_1, \mathcal{M}_2)$. It is advantageous to denote this algebraic complement as codet $A(\mathcal{M}_1, \mathcal{M}_2)$.

Theorem 1.5. (Laplace expansion theorem) *Let A be a square matrix of order n; let S be a subset of $\mathcal{N} = \{1, \ldots, n\}$. Then*

$$\det A = \sum_{\mathcal{M}} \det A(S, \mathcal{M}) \cdot \text{codet } A(S, \mathcal{M}),$$

where the summation is over all subsets $\mathcal{M} \subset \mathcal{N}$ such that card $\mathcal{M} =$ card S.

(Laplace expansion with respect to rows with indices in S.)

Remark 1.6. There is an analogous formula expanding the determinant with respect to a set of columns.

For simplicity, we denote as A_{ik} the algebraic complement of the entry a_{ik}, in the previous notation, so
$A_{ik} = \text{codet } A(\{i\}, \{k\}) = (-1)^{i+k} \det A(\mathcal{N}\backslash\{i\}, \mathcal{N}\backslash\{k\})$. We have then

$$\det A = \sum_{k=1}^{n} a_{ik} A_{ik}, \quad i = 1, \ldots, n \tag{1.3}$$

(expansion along the ith row),

$$\det A = \sum_{i=1}^{n} a_{ik} A_{ik}, \quad k = 1, \ldots, n \tag{1.4}$$

(expansion along the kth column).

Another corollary to Theorem 1.5 is:

Observation 1.7. *If a square matrix has two rows or two columns identical, its determinant is zero.*

In particular, this holds for the matrix obtained from $A = (a_{ik})$ by replacing the jth row by the ith row ($i \neq j$) of A. Expanding then the (zero) determinant of the new matrix along the jth row, we obtain

$$\sum_{k=1}^{n} a_{ik} A_{jk} = 0, \tag{1.5}$$

and this is true whenever $i \neq j$. Analogously,

$$\sum_{k=1}^{n} a_{ki} A_{kj} = 0, \tag{1.6}$$

whenever $i \neq j$.

Using the Laplace expansion theorem recurrently, we get the following:

Theorem 1.8. *Let $A = (A_{ik})$ be a block lower triangular matrix*

$$A = \begin{pmatrix} A_{11} & 0 & 0 & \ldots & 0 \\ A_{21} & A_{22} & 0 & \ldots & 0 \\ \cdot & \cdot & \cdot & \ldots & 0 \\ A_{r1} & A_{r2} & A_{r3} & \ldots & A_{rr} \end{pmatrix}$$

with r block rows and such that the diagonal blocks are square (and those above the diagonal zero). Then

$$\det A = \det A_{11} \det A_{22} \cdot \ldots \cdot \det A_{rr}.$$

Let us recall now the important *Cauchy–Binet formula* (cf. [53], Section 0.8.7).

Theorem 1.9. *Let P be an $m \times n$ matrix, Q an $n \times m$ matrix, and $m \leq n$. Let $\mathcal{M} = \{1, \ldots, m\}$, and $\mathcal{N} = \{1, \ldots, n\}$. Then*

$$\det PQ = \sum_{S} \det P(\mathcal{M}, S) \cdot \det Q(S, \mathcal{M}),$$

where the summation is over all subsets S of $\mathcal{N}$ with m elements.

For $m = n$, we obtain:

Corollary 1.10. *If P and Q are square matrices of the same order, then*

$$\det PQ = \det P \cdot \det Q.$$

We have now:

Theorem 1.11. *A matrix $A = (a_{ik})$ is nonsingular if and only if it is square and its determinant is different from zero. In addition, the inverse $A^{-1} = (\alpha_{ik})$ where*

$$\alpha_{ik} = \frac{A_{ki}}{\det A},$$

A_{ki} being the algebraic complement of a_{ki}.

Proof. We present a short proof. If A is nonsingular, then by (1.2) and Corollary 1.10,

$$\det A \cdot \det A^{-1} = 1.$$

Thus, $\det A \neq 0$. Conversely, if $\det A \neq 0$, equations (1.3) and (1.5) yield that the matrix C transposed to $(\frac{A_{ik}}{\det A})$ satisfies $AC = I$ whereas (1.4) and (1.6) imply that $CA = I$. $\qquad\square$

Remark 1.12. Corollary 1.10 implies that the product of a finite number of nonsingular matrices of the same order is again nonsingular.

Remark 1.13. Theorem 1.11 implies that for checking that the matrix C is the inverse of A, only one of the conditions $AC = I$, $CA = I$ suffices.

Let us return, for a moment, to the block lower triangular matrix in Theorem 1.8.

Theorem 1.14. *A block triangular matrix*

$$A = \begin{pmatrix} A_{11} & 0 & 0 & \ldots & 0 \\ A_{21} & A_{22} & 0 & \ldots & 0 \\ . & . & . & \ldots & 0 \\ A_{r1} & A_{r2} & A_{r3} & \ldots & A_{rr} \end{pmatrix}$$

with square diagonal blocks is nonsingular if and only if all the diagonal blocks are nonsingular. In such a case the inverse $A^{-1} = (B_{ik})$ is also lower block triangular. The diagonal blocks B_{ii} are inverses of A_{ii} and its subdiagonal blocks B_{ij}, $i > j$, can be obtained recurrently from

$$B_{ij} = -A_{ii}^{-1} \sum_{k=j}^{i-1} A_{ik} B_{kj}. \tag{1.7}$$

Proof. The condition on nonsingularity follows from Theorems 1.14 and 1.11. The blocks B_{ij} can indeed be recurrently obtained, starting with B_{21}, by increasing the difference $i - j$, since on the right-hand side of (1.7) only blocks B_{kj} with $k - j$ smaller than $i - j$ occur. Then it is easily checked that, setting all blocks B_{ik} for $i < k$ as zero blocks, and B_{ii} as A_{ii}^{-1}, all conditions for $AB = I$ are fulfilled. By Remark 1.13, (B_{ik}) is indeed A^{-1}. $\square$

Remark 1.15. This theorem applies, of course, also to the simplest case when the blocks A_{ik} are entries of the lower triangular matrix (a_{ik}). An analogous result on inverting upper triangular matrices, or upper block triangular matrices, follows by transposing the matrix and using Observation 1.3.

Corollary 1.16. *The class of lower triangular matrices of the same order is closed with respect to addition, scalar multiplication, and matrix multiplication as well as, for nonsingular matrices, to inversion. The same is true for upper triangular matrices, and also for diagonal matrices.*

As we saw, triangular matrices can be inverted rather simply. This enables us to invert matrices that allow factorization into triangular matrices. This is possible in the following case.

A square matrix A of order n is called *strongly nonsingular* if all its principal minors $\det A(\mathcal{N}_k, \mathcal{N}_k)$, $k = 1, \ldots, n$, $\mathcal{N}_k = \{1, \ldots, k\}$ are different from zero.

Theorem 1.17. *Let A be a square matrix. Then the following are equivalent:*

1. A is strongly nonsingular.

2. A has an LU-decomposition, i.e., there exist a nonsingular lower triangular matrix L and a nonsingular upper triangular matrix U such that $A = LU$.

The condition 2 can be formulated in a stronger form.

$A = BDC$, where B is a lower triangular matrix with ones on the diagonal, C is an upper triangular matrix with ones on the diagonal and D is a nonsingular diagonal matrix. This factorization is uniquely determined. The diagonal entries d_k of D are

$$d_1 = A(\{1\}, \{1\}), d_k = \frac{\det A(\mathcal{N}_k, \mathcal{N}_k)}{\det A(\mathcal{N}_{k-1}, \mathcal{N}_{k-1})}, k = 2, \ldots, n.$$

The proof is left to the reader.

Let now

$$A = \begin{pmatrix} A_{11} & A_{12} \\ A_{21} & A_{22} \end{pmatrix}$$

be a block matrix in which A_{11} is nonsingular. We then call the matrix

$$A_{22} - A_{21} A_{11}^{-1} A_{12}$$

the *Schur complement* of the submatrix A_{11} in A and denote it by $[A/A_{11}]$. Here, the matrix A_{22} need not be square.

Theorem 1.18. *If the matrix*

$$A = \begin{pmatrix} A_{11} & A_{12} \\ A_{21} & A_{22} \end{pmatrix}$$

is square and A_{11} is nonsingular, then the matrix A is nonsingular if and only if the Schur complement $[A/A_{11}]$ is nonsingular. We have then

$$\det A = \det A_{11} \det[A/A_{11}],$$

and if the inverse

$$A^{-1} = \begin{pmatrix} B_{11} & B_{12} \\ B_{21} & B_{22} \end{pmatrix}$$

is written in the same block form, then

$$[A/A_{11}] = B_{22}^{-1}.$$

The proof is simple.

Observe that the system of m linear equations with n unknowns

$$a_{11}x_1 + a_{12}x_2 + \cdots + a_{1n}x_n = b_1,$$
$$a_{21}x_1 + a_{22}x_2 + \cdots + a_{2n}x_n = b_2,$$
$$\cdots$$
$$a_{m1}x_1 + a_{m2}x_2 + \cdots + a_{mn}x_n = b_n$$

can be written in the form
$$Ax = b, \tag{1.8}$$

where the $m \times n$ matrix $A = (a_{ik})$ is the *matrix of the system*, and $x = (x_1, \ldots, x_n)^T$, $b = (b_1, \ldots, b_m)^T$ are column vectors representing the *solution vector* and the *vector of the right-hand side*, respectively.

Theorem 1.19. *If the matrix of a system of n linear equations with n unknowns is nonsingular, then the system has a unique solution.*

Proof. Indeed, if A in (1.8) is nonsingular, then $x = A^{-1}b$ is the unique solution. □

Corollary 1.20. *If the system $Ax = 0$ with a square matrix A has a nonzero solution, then $\det A = 0$.*

To show that the converse also holds, we have to mention the theory on vector spaces; we suppose that the basic field is that of the real numbers, $\mathbb{R}$.

A vector space V is the set of objects called *vectors*, for which two operations are defined: *addition* denoted by $+$ and (sometimes called *scalar*) *multiplication by a number* (in our case, from $\mathbb{R}$) denoted, for the moment, by $\circ$.

The following properties have to be fulfilled.

(V1) $u + v = v + u$ for all u, v in V;
(V2) $(u + v) + w = u + (v + w)$ for all u, v and w in V;
(V3) There exists a vector $0 \in V$ (*the zero vector*) such that $u + 0 = u$ for all $u \in V$;
(V4) If $u \in V$, then there is in V a vector $-u$ (*the opposite vector*) such that $u + (-u) = 0$;
(V5) $\alpha \circ (u + v) = \alpha \circ u + \alpha \circ v$ for all $u \in V$, $v \in V$, and $\alpha \in \mathbb{R}$;
(V6) $(\alpha + \beta) \circ u = \alpha \circ u + \beta \circ u$ for all $u \in V$ and all α, β in $\mathbb{R}$;
(V7) $(\alpha\beta) \circ u = \alpha \circ (\beta \circ u)$ for all $u \in V$ and all α, β in $\mathbb{R}$;
(V8) $-u = (-1) \circ u$ for all $u \in V$.

Here, the most important case (serving also as an example in the nearest sequel) is the *n-dimensional arithmetic vector space*, namely the set $\mathbb{R}^n$ of all real column vectors $(a_1, \ldots, a_n)^T$ with addition as defined above for matrices $n \times 1$ and multiplication by a number as scalar multiplication for matrices. Analogously, if $\mathbb{C}^n$ is the set of all complex column vectors with such addition and scalar multiplication, the scalars are complex.

A finite system of vectors $u_1, u_2, \ldots, u_s$ in V is called *linearly dependent*, if there exist numbers $\alpha_1, \alpha_2, \ldots, \alpha_s$ in $\mathbb{R}$ not all equal to zero and such that

$$\alpha_1 \circ u_1 + \alpha_2 \circ u_2 + \cdots + \alpha_s \circ u_s = 0.$$

Otherwise, the system is called *linearly independent*.

In the example of the vector space $\mathbb{R}^2$, the system of vectors

$$\begin{pmatrix} 1 \\ 0 \end{pmatrix}, \quad \begin{pmatrix} 0 \\ 1 \end{pmatrix}, \quad \begin{pmatrix} a_1 \\ a_2 \end{pmatrix}$$

is linearly dependent, since

$$a_1 \begin{pmatrix} 1 \\ 0 \end{pmatrix} + a_2 \begin{pmatrix} 0 \\ 1 \end{pmatrix} + (-1) \begin{pmatrix} a_1 \\ a_2 \end{pmatrix} = \begin{pmatrix} 0 \\ 0 \end{pmatrix},$$

and the third coefficient -1 is always different from zero. The system

$$\begin{pmatrix} 1 \\ 0 \end{pmatrix}, \quad \begin{pmatrix} 0 \\ 1 \end{pmatrix}$$

is linearly independent, since if

$$\alpha_1 \begin{pmatrix} 1 \\ 0 \end{pmatrix} + \alpha_2 \begin{pmatrix} 0 \\ 1 \end{pmatrix} = \begin{pmatrix} 0 \\ 0 \end{pmatrix}$$

holds, then by comparing the first entries on the left and on the right $\alpha_1 = 0$, from the second entries $\alpha_2 = 0$ as well; thus, no such nonzero pair of numbers α_1, α_2 exists.

If $u_1, u_2, \ldots, u_s$ is a system of vectors in V and v a vector in V, we say that v is *linearly dependent on* (or, equivalently, is a *linear combination of*) $u_1, u_2, \ldots, u_s$, if there exist numbers $\alpha_1, \alpha_2, \ldots, \alpha_s$ in $\mathbb{R}$ such that $v = \alpha_1 \circ u_1 + \alpha_2 \circ u_2 + \cdots + \alpha_s \circ u_s$.

A vector space has *finite dimension* if there exists a nonnegative integer m such that every system of vectors in V with more than m vectors is linearly dependent. The *dimension* of such V is then the smallest of such numbers m; in other words, it is a number n with the property that there is a system of n linearly independent vectors in V, but every system having more than n vectors is already linearly dependent. Such a system of n linearly independent vectors of an n-dimensional vector space V is called the *basis* of V.

The arithmetic vector space $\mathbb{R}^n$ then has dimension n since the system $e_1 = (1, 0, \ldots, 0)^T, e_2 = (0, 1, \ldots, 0)^T, \ldots, e_n = (0, 0, \ldots, 1)^T$ is a basis of $\mathbb{R}^n$.

Observation 1.21. *The set $\mathbb{R}^{m \times n}$ of real $m \times n$ matrices is also a vector space; it has dimension mn.*

If V_1 is a nonempty subset in a vector space V which is closed with respect to the operations of addition and scalar multiplication in V, then we say that V_1 is a *linear subspace* of V. It is clear that the intersection of linear subspaces of V is again a linear subspace of V. In this sense, the set (0) is in fact a linear subspace contained in all linear subspaces of V.

If S is some set of vectors of a finite-dimensional vector space V, then the linear subspace of V of smallest dimension that contains the set S is called the *linear hull* of S and its dimension (necessarily finite) is called the *rank* of S.

We are now able to present, without proof, an important statement about the rank of a matrix.

Theorem 1.22. *Let A be an $m \times n$ matrix. Then the rank of the system of the columns (as vectors) of A is the same as the rank of the system of the rows (as vectors) of A. This common number $r(A)$, called the rank of the matrix A, is equal to the maximum order of all nonsingular submatrices of A. (If A is the zero matrix, thus containing no nonsingular submatrix, then $r(A) = 0$.)*

We can now complete Theorem 1.11 and Corollary 1.20.

Theorem 1.23. *A square matrix A is singular if and only if there exists a nonzero vector x for which $Ax = 0$.*

Proof. The "if" part is in Corollary 1.20. Let now A of order n be singular. By Theorem 1.22, $r(A) \leq n - 1$ so that the system of columns $A_1, A_2, \ldots, A_n$ of A is linearly dependent. If $x_1, x_2, \ldots, x_n$ are those (not all zero) coefficients for which

$$x_1 A_1 + x_2 A_2 + \cdots + x_n A_n = 0,$$

then indeed $Ax = 0$ for $x = (x_1, x_2, \ldots, x_n)^T$, $x \neq 0$.   □

The rank function enjoys important properties. We list some:

Theorem 1.24. *We have:*

1. For any matrix A,

$$r(A^T) = r(A).$$

2. If the matrices A and B have the same type, then

$$r(A + B) \leq r(A) + r(B).$$

3. If the matrices A and B can be multiplied, then

$$r(AB) \leq \min(r(A), r(B)).$$

4. If A (resp., B) is nonsingular, then $r(AB) = r(B)$ (resp., $r(AB) = r(A)$).
5. If a matrix A has rank one, then there exist column vectors x and y such that $A = xy^T$.

We leave the proof to the reader; let us only remark that the following formula for the determinant of the sum of two square matrices of the same order n can be used,

$$\det(A + B) = \sum_{\mathcal{M}_i, \mathcal{M}_j} \det A(\mathcal{M}_i, \mathcal{M}_j) \cdot \operatorname{codet} B(\mathcal{M}_i, \mathcal{M}_j),$$

where the summation is taken over all pairs $\mathcal{M}_i$, $\mathcal{M}_j$ of subsets of $\mathcal{N} = \{1, \ldots, n\}$ that satisfy $\operatorname{card} \mathcal{M}_i = \operatorname{card} \mathcal{M}_j$.

For square matrices, the following important notions have to be mentioned.

Let A be a square matrix of order n. A nonzero column vector x is called the *eigenvector* of A if $Ax = \lambda x$ for some number (scalar) λ. This number λ is called the *eigenvalue* of A corresponding to the eigenvector x.

Theorem 1.25. *A necessary and sufficient condition that a number λ is an eigenvalue of a matrix A is that the matrix $A - \lambda I$ is singular, i.e., that*

$$\det(A - \lambda I) = 0.$$

This formula is equivalent to

$$(-\lambda)^n + c_1(-\lambda)^{n-1} + \cdots + c_{n-1}(-\lambda) + c_n = 0, \tag{1.9}$$

where c_k is the sum of all principal minors of A of order k,

$$c_k = \sum_{M \subset \mathcal{N}, \, \text{card } M = k} \det A(\mathcal{M}, \mathcal{M}), \quad \mathcal{N} = \{1, \dots, n\}.$$

The polynomial on the left-hand side of (1.9) is called the *characteristic polynomial* of the matrix A. It has degree n.

We have thus:

Theorem 1.26. *A square complex matrix $A = (a_{ik})$ of order n has n eigenvalues (some may coincide). These are all the roots of the characteristic polynomial of A. If we denote them as $\lambda_1, \dots, \lambda_n$, then*

$$\sum_{i=1}^{n} \lambda_i = \sum_{i=1}^{n} a_{ii}, \tag{1.10}$$

$$\lambda_1 \lambda_2 \cdot \ldots \cdot \lambda_n = \det A.$$

The number $\sum_{i=1}^{n} a_{ii}$ is called the *trace* of the matrix A. We denote it by $\operatorname{tr} A$. By (1.10), $\operatorname{tr} A$ is the sum of all eigenvalues of A.

Remark 1.27. A real square matrix need not have real eigenvalues, but as its characteristic polynomial is real, the nonreal eigenvalues occur in complex conjugate pairs.

We say that a square matrix B is *similar* to the matrix A if there exists a nonsingular matrix P such that $B = PAP^{-1}$. The relation $\sim$ of similarity is *reflexive*, i.e. $A \sim A$, *symmetric*, i.e. if $A \sim B$, then $B \sim A$, and *transitive* , i.e. if $A \sim B$ and $B \sim C$, then $A \sim C$. Therefore, the set of square matrices of the same order splits into classes of matrices, each class containing mutually similar matrices.

The problem of what these classes look like is answered in the following theorem whose proof is omitted (cf. [53], Section 3.1). We say that a matrix is *in the Jordan normal form* if it is block diagonal with *Jordan blocks* of the form

$$J_k(\sigma) = \begin{pmatrix} \sigma & 1 & 0 & \dots & 0 & 0 \\ 0 & \sigma & 1 & \dots & 0 & 0 \\ 0 & 0 & \sigma & \dots & 0 & 0 \\ \cdot & \cdot & \cdot & \cdot & \cdot & \cdot \\ 0 & 0 & 0 & \dots & \sigma & 1 \\ 0 & 0 & 0 & \dots & 0 & \sigma \end{pmatrix}.$$

Theorem 1.28. *Every real or complex square matrix is in the complex field similar to a matrix in the Jordan normal form. Moreover, if two matrices have the same Jordan normal form apart from the ordering of the diagonal blocks, then they are similar.*

Theorem 1.29. *A real or complex square matrix is nonsingular if and only if all its eigenvalues are different from zero. In such case, the inverse has eigenvalues reciprocal to the eigenvalues of the matrix.*

Given a square matrix A and a polynomial $f(x) = a_m x^m + a_{m-1} x^{m-1} + \cdots + a_1 x + a_0$, we speak about the *polynomial $f(A)$ in the matrix A* defined as follows: $f(A) = a_m A^m + a_{m-1} A^{m-1} + \cdots + a_1 A + a_0 I$.

Theorem 1.30. *If λ is an eigenvalue of A with eigenvector x, then $f(\lambda)$ is an eigenvalue of $f(A)$ with eigenvector x.*

Similar matrices A and B satisfying $B = PAP^{-1}$ then have the property that for every polynomial $f(x)$, $f(B) = Pf(A)P^{-1}$.

To show the importance of Theorem 1.28, let us first introduce the notion of the spectral radius of a square matrix A. If $\lambda_1, \ldots, \lambda_n$ are all eigenvalues of A, then the *spectral radius $\varrho(A)$* of A is

$$\varrho(A) = \max_{i=1,\ldots,n} |\lambda_i|.$$

Theorem 1.31. *Let A be a square real or complex matrix. If the spectral radius $\varrho(A)$ is less than 1, then*

1. $\lim_{k \to \infty} A^k = 0$.
2. The series

$$I + A + A^2 + \cdots$$

converges to $(I - A)^{-1}$.

Proof. By Theorem 1.28, the matrix is similar to a matrix in the Jordan normal form J:

$$A = PJP^{-1}.$$

Since A and J have the same eigenvalues, the spectral radius $\varrho(J) < 1$. It is easily seen that then the powers of all Jordan blocks of J converge to zero. Thus $\lim_{k \to \infty} J^k = 0$, and since $A^k = PJ^k P^{-1}$, $\lim_{k \to \infty} A^k = 0$ as well.

Let us prove assertion 2. By Theorem 1.30, $I - A$ has eigenvalues $1 - \lambda_k$ where λ_ks are eigenvalues of A. Since $|\lambda_k| < 1$ for all k, $I - A$ is nonsingular by Theorem 1.29.

Now,

$$I + A + A^2 + \cdots + A^k = (I - A^{k+1})(I - A)^{-1},$$

and A^{k+1} converges to 0 by assertion 1. The result follows. $\qquad\square$

1.2 Norms, basic numerical linear algebra

Very often, in particular in applications, we have to add to the vector structure in a vector space notions allowing us to measure the vectors and the related objects.

Suppose we have a vector n-dimensional space V_n, real or more generally, complex. As shown later, the usual approaches to how to assign to a vector x its magnitude can be embraced by the general definition of a norm.

A *norm* in V_n is a function g that assigns to every vector $x \in V_n$ a *non-negative* number $g(x)$ and enjoys the following properties.

N1. $g(x + y) \leq g(x) + g(y)$ for all $x \in V_n$ and $y \in V_n$;
N2. $g(\lambda \circ x) = |\lambda| g(x)$ for all $x \in V_n$ and all scalars λ;
N3. $g(x) = 0$ only if $x = 0$.

In the example of the vector space $\mathbb{C}^n$ the following are the most useful norms assigned to a column vector $x = (x_1, \ldots, x_n)^T$:

$$g_1(x) = \sum_{i=1}^{n} |x_i|;$$

$$g_2(x) = \sqrt{\sum_{i=1}^{n} |x_i|^2}; \tag{1.11}$$

$$g_3(x) = \max_i |x_i|.$$

The first norm is sometimes called the *octahedric norm* or l_1-*norm*, the second *Euclidean norm* or l_2-*norm*, and the third *max-norm*, *cubic norm*, or l_∞-*norm*. All these norms satisfy the above properties N1 to N3. We should mention here that the norm $g_2(.)$ is usually denoted by $||.||$ in the sequel.

As we know from Observation 1.21, the sets $\mathbb{R}^{m \times n}$ and $\mathbb{C}^{m \times n}$ of real or complex $m \times n$ matrices form vector spaces of dimension mn. In these spaces, usually the *Frobenius norm* is used: If $A = (a_{ik})$, then this norm is defined analogously to g_2 above as

$$N(A) = \sqrt{\sum_{i=1}^{m} \sum_{k=1}^{n} |a_{ik}|^2}.$$

However, theoretically the most important norms in $\mathbb{R}^{m \times n}$ and $\mathbb{C}^{m \times n}$ are the matrix norms *subordinate to the vector norms* defined below. From now on we restrict ourselves to the case of square matrices.

Let g be a vector norm in, say, $\mathbb{C}^n$. If A is in $\mathbb{C}^{n \times n}$, then we define

$$g(A) = \sup\{g(Ax) \mid x \in \mathbb{C}^n,\ g(x) = 1\},$$

or equivalently,

$$g(A) = \max_{x \neq 0, \, x \in \mathbb{C}^n} \frac{g(Ax)}{g(x)}. \tag{1.12}$$

It can be proved that (1.12) is indeed a norm on the space of matrices. Moreover, we even have for the product of matrices

$$g(AB) \leq g(A)g(B),$$

and for the identity matrix

$$g(I) = 1.$$

Remark 1.32. The last formula shows that the Frobenius norm is not a subordinate norm for $n > 1$ since $N(I) = \sqrt{n}$.

In the case of the g_1-norm, the corresponding matrix norm of the matrix $A = (a_{ik})$ is

$$g_1(A) = \max_k \sum_i |a_{ik}|,$$

for the g_3-norm,

$$g_3(A) = \max_i \sum_k |a_{ik}|.$$

The matrix norm corresponding to $g_2(A)$ is used in Section 3.

There is an important relationship between subordinate matrix norms and the spectral radius.

Theorem 1.33. *For any subordinate norm g and any square matrix A, we have*

$$\varrho(A) \leq g(A).$$

Proof. If $\varrho(A) = |\lambda_i|$ for an eigenvalue λ_i of A, let y be a corresponding eigenvector. By (1.12), $g(A) \geq \frac{g(Ay)}{g(y)}$ and the right-hand side is $\varrho(A)$.   □

Let us mention the notion of *duality* which plays an important role in linear algebra and linear programming as well as in many other fields. In the most general case, two vector spaces V and V' over the same field F are called *dual* if there exists a *bilinear form* $\langle x, x' \rangle$, a function $V \times V' \to F$ satisfying besides bilinearity:

B1. $\langle x_1 + x_2, x' \rangle = \langle x_1, x' \rangle + \langle x_2, x' \rangle$ for all $x_1 \in V,\ x_2 \in V,\ x' \in V'$;
B2. $\langle \lambda x, x' \rangle = \lambda \langle x, x' \rangle$ for all $x \in V,\ x' \in V'$ and $\lambda \in F$;
B3. $\langle x, x_1' + x_2' \rangle = \langle x, x_1' \rangle + \langle x, x_2' \rangle$ for all $x \in V,\ x_1' \in V'\ x_2' \in V'$;
B4. $\langle x, \mu x' \rangle = \mu \langle x, x' \rangle$ for all $x \in V,\ x' \in V'$ and $\mu \in F$;

the two conditions:

1. For every nonzero vector $x \in V$ there exists a vector $x' \in V'$ such that $\langle x, x' \rangle \neq 0$;
2. For every nonzero vector $x' \in V'$ there exists a vector $x \in V$ such that $\langle x, x' \rangle \neq 0$.

It can be shown that both spaces V and V' have the same dimension and, in addition, there exist so-called *dual bases*; for the finite-dimensional case of dimensions n, these are bases $e_1, \ldots, e_n$ in V and $e'_1, \ldots, e'_n$ in V' for which $\langle e_i, e'_j \rangle = \delta_{ij}$ (called the *Kronecker delta*; i.e., $\delta_{ij} = 1$ if $i = j$, $\delta_{ij} = 0$ if $i \neq j$).

For example, if V is the vector space of column vectors, V' is the vector space of row vectors of the same dimension with respect to the bilinear form $\langle x, x' \rangle = x'x$, the product of the vectors x' and x.

However, V' can then also be the set of *linear functions* on V, i.e., functions $f(x) : V \to F$ satisfying $f(x + y) = f(x) + f(y)$ for all $x \in V$, $y \in V$, and $f(\lambda x) = \lambda f(x)$ for all $x \in V$ and all $\lambda \in F$. These functions can again be added and multiplied by scalars, as the bilinear form can simply serve $\langle x, f \rangle = f(x)$.

Let us return now to solving linear systems. As we observed in (1.8), such a system has the form $Ax = b$. The general criterion, which is, however, rather theoretical, is due to Frobenius.

Theorem 1.34. *The linear system*

$$Ax = b$$

has a solution if and only if both matrices A and the block matrix $(A\ b)$ have the same rank. This is always true if A has linearly independent rows.

To present a more practical approach to the problem, observe first that permuting the equations of the system does not essentially change the problem. Algebraically, it means multiplication of (1.8) from the left by a *permutation matrix* P, i.e., a square matrix that has in each row as well as in each column only one entry different from zero, always equal to one. Of course, such a matrix satisfies

$$PP^T = I. \tag{1.13}$$

Similarly, we can permute the columns of A and the rows of x, multiplying A from the right by a permutation matrix Q; i.e., we insert the matrix QQ^T between A and x:

$$(AQ)(Q^T x) = b.$$

Thus, in general, systems of linear equations with matrices A and PAQ, where P and Q are permutation matrices, are trivially equivalent.

Another observation is that a system of linear equations with a square upper triangular nonsingular matrix of the system is easily solved: we compute the last unknown from the last equation, substitute the result into the last but one equation to obtain the last but one unknown, etc., until the whole vector of the solution is found. This method is called the *backward substitution*.

A similar procedure can be applied to more general systems whose matrix has the so-called *row echelon form*[1].

---

[1] The first column of zeros need not always be present.

$$A = \begin{pmatrix} 0 & A_{11} & A_{12} & \dots & A_{1r} & \dots & A_{1s} \\ 0 & 0 & A_{22} & \dots & A_{2r} & \dots & A_{2s} \\ . & . & . & \dots\dots\dots & . \\ 0 & 0 & 0 & \dots & A_{rr} & \dots & A_{rs} \\ 0 & 0 & 0 & \dots & 0 & \dots & 0 \end{pmatrix},$$

where r is the rank and A_{kk} are row vectors with the first coordinate equal to one.

One can show that every matrix can be brought to such form by multiplication from the left by a nonsingular matrix, i.e., by performing row operations only. These operations can even be done by stepwise performing *elementary row operations*, which are:

1. Multiplication of a row by a nonzero number;
2. Adding a row multiplied by a nonzero number to another row;
3. Exchanging two rows.

If we perform these row operations on the block matrix $(A\ b)$, until A reaches such form, it is easy to decide whether the system has a solution and then find all solutions.

The algorithm that transforms the matrix by row operations into the row echelon form can be, at least theoretically, done by the *Gaussian elimination method*. One finds the first nonzero column, finds the first nonzero entry in it, by operation 3 puts it into the first place, changes it by operation 1 into one, eliminates using operation 2 all the remaining nonzero entries in this column, and continues with the submatrix left after removing the first row, in the same way, until no row is left.

It might, however, happen that the first nonzero entry (in the first step or in further steps) is very small in modulus. Then it is better to choose another entry in the relevant column which has a bigger modulus. This entry is then called the *pivot* in the relevant step.

Observe that operation 1 above corresponds to multiplication from the left by a nonsingular diagonal matrix differing from the identity by just one diagonal entry. Operation 2 corresponds to multiplication from the left by a matrix of the form $I + \alpha E_{ik}$, where E_{ik} is the matrix with just one entry 1 in the position (i,k) and zeros elsewhere; here, $i \neq k$. Operation 3 finally corresponds to multiplication from the left by a permutation matrix obtained from the identity by switching just two rows. Thus, altogether, we have:

Theorem 1.35. *Every system $Ax = b$ can be transformed into an equivalent system $\hat{A}\hat{x} = \hat{b}$ in which the matrix $(\hat{A}\ \hat{b})$ has the row-echelon form by multiplication from the left by a nonsingular matrix.*

Remark 1.36. If the matrix A of such a system is strongly nonsingular, i.e., if it has an LU-decomposition from Theorem 1.17, we can use the pivots (1,1), (2,2) etc., and obtain the echelon form as an upper-triangular matrix with ones on the diagonal. The nonsingular matrix by which we multiply the system is then the matrix L^{-1} where $A = LU$ is the decomposition.

One can also use the more general Gaussian block elimination method.

Theorem 1.37. *Let the system $Ax = b$ be in the block form*

$$A_{11}x_1 + A_{12}x_2 = b_1,$$
$$A_{21}x_1 + A_{22}x_2 = b_2,$$

where x_1, x_2 are vectors.

If A_{11} is nonsingular, then this system is equivalent to the system

$$A_{11}x_1 + A_{12}x_2 = b_1,$$
$$(A_{22} - A_{21}A_{11}^{-1}A_{12})x_2 = b_2 - A_{21}A_{11}^{-1}b_1.$$

Proof. We perform one step of the Gaussian elimination by multiplying the first block equation by A_{11}^{-1} from the left and subtracting it multiplied by A_{21} from the left of the second equation. Then the resulting system has a block echelon form (we left there the block diagonal coefficient matrices).   □

Remark 1.38. In this theorem, the role of the Schur complement $[A/A_{11}] = A_{22} - A_{21}A_{11}^{-1}A_{12}$ for elimination is recognized.

In numerical linear algebra, the so-called *iterative methods* nowadays play a very important role, for instance for solving large systems of linear equations.

Let us describe the simplest *Jacobi method*. Write the given system of linear equations with a square matrix in the form

$$(I - A)x = b. \tag{1.14}$$

We choose an *initial vector* x_0 and set

$$
\begin{aligned}
x_1 &= Ax_0 + b, \\
x_2 &= Ax_1 + b, \\
&\cdots \\
x_{k+1} &= Ax_k + b,
\end{aligned}
\tag{1.15}
$$

etc.

If the sequence of the vectors $\{x_k\}$ converges to a vector $\hat{x}$, then it is clear that $\hat{x}$ is a solution of (1.14).

Theorem 1.39. *Let the spectral radius $\varrho(A)$ of the matrix A in (1.14) satisfy*

$$\varrho(A) < 1. \tag{1.16}$$

Then the sequence of vectors formed in (1.15) converges for any initial vector x_0 to the solution of (1.14) which is unique.

A sufficient condition for (1.16) is that for some norm g subordinate to a vector norm,

$$g(A) < 1.$$

Proof. By induction, the formula

$$x_k = A^k x_0 + (I + A + A^2 + \cdots + A^{k-1})b \text{ for } k = 1, 2, \ldots$$

is easily proved. By Theorem 1.31, x_k converges to $(I - A)^{-1}b$, which is the unique solution of (1.14). The last assertion follows from Theorem 1.33.   □

1.3 Symmetric matrices

In this section, we pay attention to a specialized real vector space called *Euclidean vector space* in which magnitude (length) of a vector is defined using the so-called *inner product* of two vectors.

In a few cases, we also consider complex vector spaces; the interested reader can find the related theory of the *unitary vector space* in [35].

A real finite-dimensional vector space E is called a *Euclidean vector space* if a function $\langle x, y \rangle : E \times E \to \mathbb{R}$ is given that satisfies:

E1. $\langle x, y \rangle = \langle y, x \rangle$ for all $x \in E$, $y \in E$;
E2. $\langle x_1 + x_2, y \rangle = \langle x_1, y \rangle + \langle x_2, y \rangle$ for all $x_1 \in E$, $x_2 \in E$, and $y \in E$;
E3. $\langle \alpha x, y \rangle = \alpha \langle x, y \rangle$ for all $x \in E$, $y \in E$, and all real α;
E4. $\langle x, x \rangle \geq 0$ for all $x \in E$, with equality if and only if $x = 0$.

The property E4 enables us to define the *length* $\|x\|$ of the vector x as $\sqrt{\langle x, x \rangle}$. A vector is called a *unit vector* if its length is one. Vectors x and y are *orthogonal* if $\langle x, y \rangle = 0$. A system $u_1, \ldots, u_m$ of vectors in E is called *orthonormal* if $\langle u_i, u_j \rangle = \delta_{ij}$, the Kronecker delta.

It is easily proved that every orthonormal system of vectors is linearly independent. If the number of vectors in such a system is equal to the dimension of E, it is called an *orthonormal basis* of E.

The real vector space $\mathbb{R}^n$ of column vectors will become a Euclidean space if the inner product of the vectors $x = (x_1, \ldots, x_n)^T$ and $y = (y_1, \ldots, y_n)^T$ is defined as

$$\langle x, y \rangle = x_1 y_1 + \cdots + x_n y_n.$$

An example of an orthonormal basis is the system $e_1 = (1, 0, \ldots, 0)^T$, $e_2 = (0, 1, \ldots, 0)^T$, ..., $e_n = (0, 0, \ldots, 1)^T$.

Theorem 1.40. *If $A = (a_{ik})$ is in $\mathbb{R}^{n \times n}$, then for $x \in \mathbb{R}^n$ and $y \in \mathbb{R}^n$ we obtain*

$$\langle Ax, y \rangle = \langle x, A^T y \rangle.$$

Proof. Indeed, both sides are equal to $\sum_{i,k=1}^{n} a_{ik} x_k y_i$.   □

We now call a matrix $A = (a_{ik})$ in $\mathbb{R}^{n \times n}$ *symmetric* if $a_{ik} = a_{ki}$ for all i, k, or equivalently, if $A = A^T$. We call it *orthogonal* if $AA^T = I$. Thus:

Theorem 1.41. *The sum of two symmetric matrices in* $\mathbb{R}^{n \times n}$ *is symmetric; the product of two orthogonal matrices in* $\mathbb{R}^{n \times n}$ *is orthogonal. The identity is orthogonal and the transpose (which is equal to the inverse) of an orthogonal matrix is orthogonal.*

The following theorem on orthogonal matrices holds (see [35]).

Theorem 1.42. *Let* Q *be an* $n \times n$ *real matrix. Then the following are equivalent.*

1. Q is orthogonal.
2. For all $x \in \mathbb{R}^n$,

$$||Qx|| = ||x||.$$

3. For all $x \in \mathbb{R}^n$, $y \in \mathbb{R}^n$,

$$\langle Qx, Qy \rangle = \langle x, y \rangle.$$

4. Whenever $u_1, \ldots, u_n$ is an orthonormal basis, then $Qu_1, \ldots, Qu_n$ is an orthonormal basis as well.
5. There exists an orthonormal basis $v_1, \ldots, v_n$ such that $Qv_1, \ldots, Qv_n$ is again an orthonormal basis.

The basic theorem on symmetric matrices can be formulated as follows.

Theorem 1.43. *Let* A *be a real symmetric matrix. Then there exist an orthogonal matrix* Q *and a real diagonal matrix* D *such that* $A = QDQ^T$. *The diagonal entries of* D *are the eigenvalues of* A, *and the columns of* Q *eigenvectors of* A; *the kth column corresponds to the kth diagonal entry of* D.

Corollary 1.44. *All eigenvalues of a real symmetric matrix are real. For every real symmetric matrix there exists an orthonormal basis of* $\mathbb{R}$ *consisting of its eigenvectors.*

An important subclass of the class of real symmetric matrices is that of positive definite (resp., positive semidefinite) matrices.

A real symmetric matrix A of order n is called *positive definite* (resp., *positive semidefinite*) if for every nonzero vector $x \in \mathbb{R}^n$, the product $x^T A x$ is positive (resp., nonnegative).

In the following theorem we collect the basic *characteristic* properties of positive definite matrices. For the proof, see [35].

Theorem 1.45. *Let* $A = (a_{ik})$ *be a real symmetric matrix of order n. Then the following are equivalent.*

1. A is positive definite.
2. All principal minors of A are positive.
3. $\det A(\mathcal{N}_k, \mathcal{N}_k) > 0$ for $k = 1, \ldots, n$, where $\mathcal{N}_k = \{1, \ldots, k\}$. In other words,

$$a_{11} > 0, \ \det \begin{pmatrix} a_{11} \ a_{12} \\ a_{21} \ a_{22} \end{pmatrix} > 0, \ \det \begin{pmatrix} a_{11} \ a_{12} \ a_{13} \\ a_{21} \ a_{22} \ a_{23} \\ a_{31} \ a_{32} \ a_{33} \end{pmatrix} > 0, \ \ldots, \ \det A > 0.$$

4. *There exists a nonsingular lower triangular matrix B such that $A = BB^T$.*
5. *There exists a nonsingular matrix C such that $A = CC^T$.*
6. *The sum of all principal minors of order k is positive for $k = 1, \ldots, n$.*
7. *All eigenvalues of A are positive.*
8. *There exists an orthogonal matrix Q and a diagonal matrix D with positive diagonal entries such that $A = QDQ^T$.*

Corollary 1.46. *If A is positive definite, then A^{-1} exists and is positive definite as well.*

Remark 1.47. Observe also that the identity matrix is positive definite.

For positive semidefinite matrices, we have:

Theorem 1.48. *Let $A = (a_{ik})$ be a real symmetric matrix of order n. Then the following are equivalent.*

1. *A is positive semidefinite.*
2. *The matrix $A + \varepsilon I$ is positive definite for all $\varepsilon > 0$.*
3. *All principal minors of A are nonnegative.*
4. *There exists a square matrix C such that $A = CC^T$.*
5. *The sum of all principal minors of order k is nonnegative for $k = 1, \ldots, n$.*
6. *All eigenvalues of A are nonnegative.*
7. *There exists an orthogonal matrix Q and a diagonal matrix D with nonnegative diagonal entries such that $A = QDQ^T$.*

Corollary 1.49. *A positive semidefinite matrix is positive definite if and only if it is nonsingular.*

Corollary 1.50. *If A is positive definite and α a positive number, then αA is positive definite as well. If A and B are positive definite of the same order, then $A + B$ is positive definite; this is so, even if one of the matrices A, B is positive semidefinite.*

The expression $x^T Ax$ – in the case that A is symmetric – is called the *quadratic form* corresponding to the matrix A. It is important that the *Raleigh quotient* $\frac{x^T Ax}{x^T x}$ for $x \neq 0$ can be estimated from both sides.

Theorem 1.51. *If A is a symmetric matrix of order n with eigenvalues $\lambda_1 \geq \lambda_2 \geq \cdots \geq \lambda_n$, then*

$$\lambda_n \leq \frac{x^T Ax}{x^T x} \leq \lambda_1$$

for every nonzero vector x.

Remark 1.52. All the properties mentioned in this section hold, with appropriate changes, for the more general complex case. One defines, instead of symmetric matrices, so called *Hermitian matrices*, by $A = A^H$, where A^H means transposition and complex conjugacy. *Unitary matrices* defined by $UU^H = I$ then play the role of orthogonal matrices. It is easily shown that if A is Hermitian, $x^H A x$ is always real; positive definite is then such an Hermitian matrix for which $x^H A x > 0$ whenever x is a nonzero vector.

Now, we can fill in the gap left in the preceding section. We left open the question about the subordinate norm g_2 for matrices.

Theorem 1.53. *Let A be a (in general complex) square matrix. Then $g_2(A)$ is equal to the square root of the spectral radius $\varrho(A^H A)$. In the real case,*

$$g_2(A) = \sqrt{\varrho(A^T A)}.$$

Proof. We prove the real case only. In the notation above, and by Theorem 1.40 we get

$$\begin{aligned} g_2(Ax) &= \langle Ax, Ax \rangle^{\frac{1}{2}} \\ &= \langle A^T Ax, x \rangle^{\frac{1}{2}} \\ &\leq \langle \varrho(A^T A)x, x \rangle^{\frac{1}{2}} \\ &\leq (\varrho(A^T A))^{\frac{1}{2}} g_2(x); \end{aligned}$$

here, we also used Theorem 1.51 since $\varrho(A^T A) = \lambda_1^2$.

However, if we take an eigenvector of the symmetric positive semidefinite matrix $A^T A$ corresponding to $\varrho(A^T A)$ for x, we obtain equality.   □

For general complex matrices, even not necessarily square, the following factorization (so-called *singular value decomposition*, SVD for short) generalizes Theorem 1.43.

Theorem 1.54. *Let A be a complex $m \times n$ matrix of rank r. Then there exist unitary matrices U of order m, V of order n, and a diagonal matrix S of order r with positive diagonal entries such that*

$$A = U \begin{pmatrix} S & 0 \\ 0 & 0 \end{pmatrix} V; \tag{1.17}$$

here, the zero blocks complete the matrix to an $m \times n$ matrix. The matrix S is then determined uniquely up to the ordering of the diagonal entries.

Remark 1.55. The diagonal entries $s_1, \ldots, s_r$ of S, usually supposed ordered as $s_1 \geq s_2 \geq \cdots \geq s_r$, are called *singular values* of A.

Remark 1.56. For a real matrix A, the singular value decomposition can always be real; the matrices U and V will be orthogonal.

Concluding this section, let us notice a close relationship of the class of positive semidefinite matrices with Euclidean geometry. If $u_1, \ldots, u_m$ is a system of vectors in a Euclidean vector space, then the matrix of the inner products

$$G(u_1, \ldots, u_m) = \begin{pmatrix} \langle u_1, u_1 \rangle & \langle u_1, u_2 \rangle & \cdots & \langle u_1, u_m \rangle \\ \langle u_2, u_1 \rangle & \langle u_2, u_2 \rangle & \cdots & \langle u_2, u_m \rangle \\ & & \cdots & \\ \langle u_m, u_1 \rangle & \langle u_m, u_2 \rangle & \cdots & \langle u_m, u_m \rangle \end{pmatrix},$$

the so-called *Gram matrix* of the system, enjoys the following property.

Theorem 1.57. *The Gram matrix $G(u_1, \ldots, u_m)$ of a system of vectors in a Euclidean space is always positive semidefinite. Its rank is equal to the dimension of the linear space of the smallest dimension that contains all vectors of the system (linear hull of the system).*

Conversely, if A is an $m \times m$ positive semidefinite matrix of rank r, then there exists a Euclidean vector space of dimension r and a system of m vectors in this space the Gram matrix of which coincides with A. In addition, every linear dependence relation between the rows of A corresponds to the same linear dependence relation between the vectors of the system and conversely.

Remark 1.58. This theorem shows (in fact, it is equivalent with) that all Euclidean vector spaces of a fixed dimension are equivalent.

1.4 Generalized inverses

We complete the treatment on general matrix theory with a short section on generalized inversion.

As we know, for a nonsingular matrix A there exists a unique matrix X that satisfies $AX = I$, $XA = I$, where I is the identity matrix, namely the inverse matrix to A. Observe that this matrix X also satisfies the following relations

$$AXA = A, \tag{1.18}$$
$$XAX = X, \tag{1.19}$$
$$(AX)^T = AX, \tag{1.20}$$
$$(XA)^T = XA. \tag{1.21}$$

These relations have also meaning for matrices that are not square. Indeed, if A is $m \times n$ then for an $n \times m$ matrix X, formal conditions for multiplication of matrices are fulfilled. This observation leads to the notions of *generalized inverses* of the matrix A as matrices X that satisfy one, two, three or all conditions in (1.18) to (1.21).

Remark 1.59. In the case of complex matrices, it is useful to replace conditions (1.20) and (1.21), similarly as in Remark 1.52, by $(AX)^H = AX$ and $(XA)^H = XA$.

The existence of a generalized inverse satisfying all conditions (1.18) to (1.21) follows from the following.

Theorem 1.60. *Let* $A = USV$ *be the singular value decomposition (see (1.17)) of the* $m \times n$ *matrix* A, *where* U, V *are orthogonal (if* $A \in \mathbb{R}^{m \times n}$*) or unitary (if* $A \in \mathbb{C}^{m \times n}$*), and* S *the matrix*

$$\begin{pmatrix} S_0 & 0 \\ 0 & 0 \end{pmatrix}$$

with diagonal S_0 *having positive diagonal entries.*
Then the matrix $X = V^T \widehat{S} U^T$ *(in the complex case* $X = V^H \widehat{S} U^H$*), where*

$$\widehat{S} = \begin{pmatrix} S_0^{-1} & 0 \\ 0 & 0 \end{pmatrix}$$

is $n \times m$, *satisfies all conditions (1.18) to (1.21) (in the complex case, replaced according to Remark 1.59).*

We have, however, the following important theorem; if B is a matrix, we use the symbol B^* for the more general case of the complex conjugate transpose. In the real case, one can simply replace it by B^T.

Theorem 1.61. *In both real and complex cases, there is a unique matrix* X *that satisfies all conditions (1.18) to (1.21). In the real case,* X *is real.*

Proof. It suffices to prove the uniqueness. (We use $*$ for both the real and complex case.) By (1.18), $A^* X^* A^* = A^*$. Thus, by (1.20) and (1.21),

$$A^* AX = A^* = XAA^*; \tag{1.22}$$

similarly also

$$XX^* A^* = X = A^* X^* X. \tag{1.23}$$

Now, let both X_1 and X_2 satisfy (1.18) to (1.21). Then, by (1.22) and (1.23),

$$\begin{aligned} X_1 &= X_1 X_1^* A^* \\ &= X_1 X_1^* A^* A X_2 \\ &= X_1 A X_2 \\ &= X_1 A A^* X_2^* X_2 \\ &= A^* X_2^* X_2 \\ &= X_2. \end{aligned}$$

This unique matrix X is usually called the *Moore–Penrose inverse* (sometimes *pseudoinverse*) of A and denoted as A^+.

In the following theorem, we list the most important properties of the Moore–Penrose inverse.

Theorem 1.62. *Let A be a matrix. Then*

$$(A^+)^+ = A,$$
$$(A^*)^+ = (A^+)^*,$$
$$(AA^*)^+ = (A^+)^*A^+,$$
$$(A^*A)^+ = A^+(A^*)^+,$$
$$r(A) = r(A^+)$$
$$= r(AA^+)$$
$$= \mathrm{tr}(AA^+),$$

where $r(.)$ means the rank and $\mathrm{tr}(.)$ the trace.

If $\lambda \neq 0$ is a scalar, then $(\lambda A)^+ = \lambda^{-1}A^+$.

If U, V are unitary, then $(UAV)^+ = V^*A^+U^*$.

Corollary 1.63. *For any zero matrix, we have $0^+ = 0^T$. If the rows of A are linearly independent, then $A^+ = A^*(AA^*)^{-1}$. If the columns are linearly independent, then $A^+ = (A^*A)^{-1}A^*$. Of course, $A^+ = A^{-1}$ for a nonsingular (thus square) matrix A.*

The Moore–Penrose inverse has important applications in statistics as well as in numerical computations. If we are given a system (obtained, for instance, by repeated measuring) of m linear equations in n unknowns of the form

$$Ax = b,$$

where m is greater than n, there is usually no solution. We can then ask:

Problem. What is the best approximation x_0 of the system, i.e., for which x_0 the g_2-norm

$$\|Ax - b\|$$

attains its minimum among all vectors x in $\mathbb{R}^n$ (or, $\mathbb{C}^n$)?

The solution is given in the theorem:

Theorem 1.64. *Let A be an $m \times n$ matrix, $m \geq n$. Then the solution of the problem above is given by*

$$x_0 = A^+b,$$

where A^+ is the Moore–Penrose inverse of A.

Remark 1.65. If $m < n$, there might be more solutions of such a system of linear equations. In this case, the vector $x_0 = A^+b$ has the property that its norm $\|x_0\|$ is minimal among all solutions of the problem.

There are several ways to compute the Moore–Penrose inverse numerically. One way was already mentioned in Theorem 1.60 using the singular value decomposition. Another way is the *Greville algorithm* which constructs successively the Moore–Penrose inverses for submatrices A_k formed by the first k columns of A, $k = 1, \ldots, n$.[2] Here, a_k denotes the kth column of A. This means that $A_k = (a_1, \ldots, a_k)$, and $A = A_n$.

Theorem 1.66. *Let $A \in \mathbb{R}^{m \times n}$ (or, $\mathbb{C}^{m \times n}$). Set $A_1^+ = a_1^+$, i.e.*

$$A_1^+ = \begin{cases} (a_1^T a_1)^{-1} a_1^T & \text{if } a_1 \neq 0, \\ 0 & \text{if } a_1 = 0. \end{cases}$$

For $k = 2, \ldots, n$, define $d_k = A_{k-1}^+ a_k$, $c_k = a_k - A_{k-1} d_k$, and set

$$b_k^T = \begin{cases} c_k^+ (= (c_k^T c_k)^{-1} c_k^T) & \text{if } c_k \neq 0, \\ (1 + d_k^T d_k)^{-1} d_k A_{k-1}^+ & \text{if } c_k = 0. \end{cases}$$

Then

$$A_k^+ = \begin{pmatrix} A_{k-1}^+ - d_k b_k^T \\ b_k^T \end{pmatrix},$$

and $A^+ = A_n^+$.

Based on this theorem, we can summarize the Greville algorithm in the following form.

$c := a_1;$
if $c = 0$ then $A^+ := c^T$; else $A^+ := \frac{c^T}{c^T c}$; end
for $j = 2$ to n
$\quad d := A^+ a_j;$
$\quad c := a_j - (a_1, \ldots, a_{j-1}) d;$
$\quad$ if $c = 0$ then $b^T := \frac{d^T A^+}{1 + d^T d}$; else $b^T := \frac{c^T}{c^T c}$; end
$\quad A^+ := \begin{pmatrix} A^+ - d b^T \\ b^T \end{pmatrix};$
end

Remark 1.67. The Moore–Penrose inverse A^+ is not a continuous function of the matrix A unless the rank of A is known. This is reflected in the algorithm by deciding whether c_k is (exactly) zero. A similar problem also arises in the singular value decomposition.

---

[2]The Greville algorithm can be recommended for problems of small dimensions; otherwise, the singular value decomposition is preferable.

1.5 Nonnegative matrices, M- and P-matrices

Positivity, or more generally, nonnegativity, plays a crucial role in most parts of this book. In the present section, we always assume that the vectors and matrices are *real*.

We denote by the symbols $>$, $\geq$ or $<$, $\leq$ *componentwise comparison* of the vectors or matrices. For instance, for a matrix A, $A > 0$ means that *all* entries of A are *positive*; the matrix is called *positive*. $A \geq 0$ means nonnegativity of all entries and the matrix is called *nonnegative*.

Evidently, the sum of two or more nonnegative matrices of the same type is again nonnegative, and also the product of nonnegative matrices, if they can be multiplied, is nonnegative. Sometimes it is necessary to know whether the result is already positive. Usually, the combinatorial structure of zero and nonzero entries and not the values themselves decide. In such a case, it is useful to apply graph theory terminology. We restrict ourselves to the case of square matrices.

A *(finite) directed graph* $G = (V, E)$ consists of the *set of vertices* V and the *set of edges* E, a subset of the Cartesian product $V \times V$. This means that every edge is an ordered pair of vertices and can thus be depicted in the plane by an arc with an arrow if the vertices are depicted as points. For our purpose, V is the set $\{1, 2, \ldots, n\}$ and E the set of entries 1 of an $n \times n$ matrix $A(G)$ in the corresponding positions (i, k); if there is no edge "starting" in i and "ending" in k, the entry in the position (i, k) is zero.

We have thus assigned to a finite directed graph (usually called a *digraph*) a $(0, 1)$-matrix $A(G)$. Conversely, let $C = (c_{ik})$ be an $n \times n$ nonnegative matrix. We can assign to C a digraph $G(C) = (V, E)$ as follows: V is the set $\{1, \ldots, n\}$, and E the set of all pairs (i, k) for which c_{ik} is positive.

The graph theory terminology speaks about a *path* in G from vertex i to the vertex k if there are vertices $j_1, \ldots, j_s$ such that $(i, j_1), (j_1, j_2), \ldots, (j_s, k)$ are edges in E; $s + 1$ is then the *length* of this path. The vertices in the path need not be distinct. If they are, the path is *simple*. If i coincides with k, we speak about a *cycle*; its length is then again $s + 1$. If all the remaining vertices are distinct, the cycle is *simple*. The edges (k, k) themselves are called *loops*. The digraph is *strongly connected* if there is at least one path from any vertex to any other vertex. Further on, we show an equivalent property for matrices.

Let P be a *permutation matrix*. By (1.13), we have $PP^T = I$. If C is a square matrix and P a permutation matrix of the same order, then PCP^T is obtained from C by a *simultaneous permutation* of rows *and* columns; the diagonal entries remain diagonal. Observe that the digraph $G(PCP^T)$ differs from the digraph $G(C)$ only by different numbering of the vertices.

We say that a square matrix C is *reducible* if it has the block form

$$C = \begin{pmatrix} C_{11} & C_{12} \\ 0 & C_{22} \end{pmatrix},$$

where both matrices C_{11}, C_{22} are square of order at least one, or if it can be brought to such form by a simultaneous permutation of rows and columns.

A square matrix is called *irreducible* if it is not reducible.

This relatively complicated notion is important for nonnegative matrices and their applications (in probability theory and elsewhere). However, it has a very simple equivalent in the graph-theoretical setting.

Theorem 1.68. *A nonnegative matrix C is irreducible if and only if the digraph $G(C)$ is strongly connected.*

A more detailed view is given in the following theorem.

Theorem 1.69. *Every square nonnegative matrix can be brought by a simultaneous permutation of rows and columns to the form*

$$\begin{pmatrix} C_{11} & C_{12} & C_{13} & \dots & C_{1r} \\ 0 & C_{22} & C_{23} & \dots & C_{2r} \\ 0 & 0 & C_{33} & \dots & C_{3r} \\ . & . & . & \dots & . \\ 0 & 0 & 0 & \dots & C_{rr} \end{pmatrix}, \tag{1.24}$$

in which the diagonal blocks are irreducible (thus square) matrices.

This theorem (the proof of which is also omitted) has a counterpart in graph theory. Every finite digraph has the following structure. It consists of so-called *strong components* that are the maximal strongly connected subdigraphs; these can then be numbered in such a way that there is no edge from a vertex with a larger number of the strong component into a vertex belonging to the strong component with a smaller number.

Remark 1.70. Theorem 1.68 holds also for the case of matrices with entries in any field. The digraph of such a matrix should distinguish zero and nonzero entries only.

The importance of irreducibility for nonnegative matrices is particularly clear if we investigate powers of such a matrix. Whereas every power of a reducible matrix (1.24) is again reducible even if we add to the matrix the identity matrix, one can show that the $(n-1)$st power of $A + I$ is positive if A is an irreducible nonnegative matrix of order n.

We now state three main results of the *Perron–Frobenius theory*. For the proofs, see, e.g., [35].

Theorem 1.71. *Let A be a square nonnegative irreducible matrix of order $n > 1$. Then the spectral radius $\varrho(A)$ is a positive and simple eigenvalue of A and the corresponding eigenvector can be made positive by scalar multiplication. A nonnegative eigenvector corresponds to no other eigenvalue.*

Theorem 1.72. *Let A be a nonnegative irreducible matrix of order $n > 1$; let m be a positive integer. Then the following are equivalent.*

1. There are m eigenvalues of A having modulus $\varrho(A)$.

2. There exists a permutation matrix P such that PAP^T has the block form

$$PAP^T = \begin{pmatrix} 0 & A_{12} & 0 & \cdots & 0 \\ 0 & 0 & A_{23} & \cdots & 0 \\ \cdot & \cdot & \cdot & \cdots & \cdot \\ 0 & 0 & 0 & \cdots & A_{m-1,m} \\ A_{m1} & 0 & 0 & \cdots & 0 \end{pmatrix}$$

with square diagonal blocks, and for no permutation matrix an analogous form exists with more than m block rows.

3. The greatest common divisor of the lengths of all cycles (equivalently, simple cycles) in the graph $G(A)$ is m.

4. If $(-1)^n\lambda^n + k_{n_1}\lambda^{n_1} + k_{n_2}\lambda^{n_2} + \cdots + k_{n_s}\lambda^{n_s}$ is the characteristic polynomial of the matrix A written with nonzero terms only, $n > n_1 > \cdots > n_s \geq 0$, then m is the greatest common divisor of $n - n_1, n_1 - n_2, \ldots, n_{s-1} - n_s$.

5. m is the maximum k having the property that the rotation of the spectrum $S(A)$ (i.e., the set of the eigenvalues) of A in the complex plane by the angle $\frac{2\pi}{k}$ preserves $S(A)$.

Theorem 1.73. *Let A be a nonnegative square matrix. Then the spectral radius $\varrho(A)$ is an eigenvalue of A, and there exists a nonnegative eigenvector of A corresponding to this eigenvalue.*

Remark 1.74. The existence of a nonnegative eigenvector of a nonnegative (and nonzero) matrix can be deduced from Brouwer's fixed-point theorem, see, e.g., [7]. The mapping that maps any nonnegative unit vector x into the vector $\|Ax\|^{-1}Ax$ is continuous and maps the intersection of the unit sphere with the nonnegative orthant into itself. The existing fixed point x then corresponds to a nonnegative eigenvector, and $\|Ax\|$ is the respective eigenvalue.

There is another important class of matrices that is closely related to the previous class of nonnegative matrices.

A square matrix A is called an *M-matrix* if it has the form $kI - C$, where C is a nonnegative matrix and $k > \varrho(C)$.

Observe that every M-matrix has all off-diagonal entries non-positive. It is usual to denote the set of such matrices by Z. To characterize matrices from Z to obtain M-matrices, there exist surprisingly many possibilities. We list some:

Theorem 1.75. *Let A be a matrix in Z of order n. Then the following are equivalent.*

1. A is an M-matrix.

2. There exists a vector $x \geq 0$ such that $Ax > 0$.
3. All principal minors of A are positive.
4. The sum of all principal minors of order k is positive for $k = 1, \ldots, n$.
5. $\det A(\mathcal{N}_k, \mathcal{N}_k) > 0$ for $k = 1, \ldots, n$, where $\mathcal{N}_k = \{1, \ldots, k\}$.
6. Every real eigenvalue of A is positive.
7. The real part of every eigenvalue of A is positive.
8. A is nonsingular and A^{-1} is nonnegative.

The proof and other characteristic properties can be found in [35].

Remark 1.76. Observe the remarkable coincidence of several properties with those of positive definite matrices in Theorem 1.45. In the next theorem, we present an analogy of positive semidefinite matrices.

Theorem 1.77. *Let A be a matrix in Z of order n. Then the following are equivalent.*

1. $A + \varepsilon I$ is an M-matrix for all $\varepsilon > 0$.
2. All principal minors of A are nonnegative.
3. The sum of all principal minors of order k is nonnegative for $k = 1, \ldots, n$.
4. Every real eigenvalue of A is nonnegative.
5. The real part of every eigenvalue of A is nonnegative.

We denote matrices satisfying these conditions M_0-matrices.

Remark 1.78. As in the case of positive definite matrices, an M_0-matrix is an M-matrix if and only if it is nonsingular.

In the next theorem we list other characteristic properties of the class of real square matrices having just the property 3 from Theorem 1.75 or property 2 from Theorem 1.45, namely: All principal minors are positive. These matrices are called *P-matrices* (cf. [37], [35]).

Theorem 1.79. *Let A be a real square matrix. Then the following are equivalent.*

1. A is a P-matrix; i.e., all principal minors of A are positive.
2. Whenever D is a nonnegative diagonal matrix of the same order as A, then all principal minors of $A + D$ are different from zero.
3. For every nonzero vector $x = (x_i)$ there exists an index k such that $x_k(Ax)_k > 0$.
4. Every real eigenvalue of any principal submatrix of A is positive.
5. The implication

$$z \geq 0, \ SA^T Sz \leq 0 \ \text{implies} \ z = 0$$

holds for every diagonal matrix S with diagonal entries 1 or -1.
6. To every diagonal matrix S with diagonal entries 1 or -1 there exists a vector $x \geq 0$ such that $SASx > 0$.

We omit the proof. Let us just state three corollaries.

Corollary 1.80. *Every symmetric P-matrix is positive definite. Every P-matrix in Z is an M-matrix.*

Corollary 1.81. *If for a real square matrix A its symmetric part $\frac{1}{2}(A + A^T)$ is positive definite, then $A \in P$.*

Corollary 1.82. *If $A \in P$, then there exists a vector $x \geq 0$ such that $Ax > 0$.*

Also in this case, we can define the "topological closure" of P-matrices, namely the class of real square matrices whose all principal minors are non-negative. We call them P_0-*matrices*.

Theorem 1.83. *Let A be a real square matrix. Then the following are equivalent.*

1. A is a P_0-matrix; i.e., all principal minors of A are nonnegative.
2. Whenever D is a diagonal matrix of the same order as A with all diagonal entries positive, then all principal minors of $A + D$ are positive.
3. For every nonzero vector $x = (x_1, \ldots, x_n)^T$ there exists an index k such that $x_k(Ax)_k \geq 0$.
4. Every real eigenvalue of any principal submatrix of A is nonnegative.
5. To every diagonal matrix S with diagonal entries 1, −1, or 0, there exists a vector $y \neq 0$ such that $Sy \geq 0$, $S^2 y = y$, and $SAy \geq 0$.

As before, one can formulate some corollaries.

Corollary 1.84. *If for a real square matrix A its symmetric part $\frac{1}{2}(A + A^T)$ is positive semidefinite, then $A \in P_0$.*

Corollary 1.85. *If $A \in P_0$, then there exists a nonzero vector $x \geq 0$, such that $Ax \geq 0$.*

1.6 Examples of other special classes of matrices

A *Hankel matrix* of order n is a matrix H of the form $H = (h_{i+j})$, $i, j = 0, \ldots, n - 1$; i.e.,

$$H = \begin{pmatrix} h_0 & h_1 & h_2 & \cdots & h_{n-1} \\ h_1 & h_2 & \cdots & h_{n-1} & h_n \\ h_2 & \cdots & & & \\ \cdots & & \cdots & & \cdots \\ h_{n-1} & & & \cdots & h_{2n-3} & h_{2n-2} \end{pmatrix}.$$

Its entries h_k can be real or complex. Let $\mathcal{H}_n$ denote the class of all $n \times n$ Hankel matrices. Evidently, $\mathcal{H}_n$ is a linear vector space (complex or real) of dimension $2n-1$. It is also clear that an $n \times n$ Hankel matrix has rank one if and only if it is either of the form $\gamma(t^{i+k})$ for γ and t fixed (in general, complex), or if it has a single nonzero entry in the lower-right corner. Hankel matrices play an important role in approximations, investigation of polynomials, etc.

A closely related class is that of *Toeplitz matrices*. These are (if of order n) matrices of the form $T = (t_{i-k})$, i.e.,

$$\begin{pmatrix} t_0 & t_1 & t_2 & \cdots & t_{n-2} & t_{n-1} \\ t_{-1} & t_0 & t_1 & \cdots & t_{n-3} & t_{n-2} \\ t_{-2} & t_{-1} & t_0 & \cdots & t_{n-4} & t_{n-3} \\ \cdots & \cdots & \cdots & \cdots & \cdots & \cdots \\ t_{-n+1} & t_{-n+2} & t_{-n+3} & \cdots & t_{-1} & t_0 \end{pmatrix}.$$

Its entries t_k can again be real or complex. The class $\mathcal{T}_n$ of all $n \times n$ Toeplitz matrices is also a vector space of dimension $2n - 1$.

An important subclass of $\mathcal{T}_n$ is that of *circulant matrices*, i.e., matrices of the form

$$\begin{pmatrix} c_0 & c_1 & c_2 & \cdots & c_{n-2} & c_{n-1} \\ c_{n-1} & c_0 & c_1 & \cdots & c_{n-3} & c_{n-2} \\ c_{n-2} & c_{n-1} & c_0 & \cdots & c_{n-4} & c_{n-3} \\ \cdots & \cdots & \cdots & \cdots & \cdots & \cdots \\ c_1 & c_2 & c_3 & \cdots & c_{n-1} & c_0 \end{pmatrix}.$$

These matrices also form a vector space, this time of dimension n. Observe that the matrix above can be expressed as

$$c_0 I + c_1 S + c_2 S^2 + \cdots + c_{n-1} S^{n-1},$$

where S is the permutation matrix

$$S = \begin{pmatrix} 0 & 1 & 0 & \cdots & 0 \\ 0 & 0 & 1 & \cdots & 0 \\ \cdots & \cdots & \cdots & \cdots & \cdots \\ 0 & 0 & 0 & \cdots & 1 \\ 1 & 0 & 0 & \cdots & 0 \end{pmatrix}.$$

By Theorem 1.30, the spectral properties of a circulant matrix follow easily from the spectral properties of the matrix S. It is easily checked that the eigenvalues of S are exactly all nth roots of 1; if ε is such a root, then $(1, \varepsilon, \varepsilon^2, \ldots, \varepsilon^{n-1})^T$ is the corresponding eigenvector.

A real $m \times n$ matrix $C = (c_{ik})$ is called (cf. [52], [23]) a *Monge matrix* if it satisfies

$$c_{ik} + c_{jl} \le c_{il} + c_{jk} \quad \text{for all} \quad i, j, k, l, \quad i < j, \ k < l. \tag{1.25}$$

Monge matrices play an important role in assignment and transportation problems. In the case of such a matrix there is a simple solution. A survey of Monge matrices and their applications is given in [23].

In our considerations, it is simpler to consider the class of matrices called in accordance with [36] *anti-Monge*. This class is defined by inequalities (1.26) analogous to (1.25) but with opposite signs of inequalities:

$$c_{ik} + c_{jl} \geq c_{il} + c_{jk} \quad \text{for all} \quad i,j,k,l, \quad i < j, \ k < l. \tag{1.26}$$

One can show easily that among $\binom{m}{2}\binom{n}{2}$ inequalities (1.26) only $(m-1)(n-1)$ are relevant, namely those for which $j = i+1$ and $l = k+1$, $i = 1,\ldots,m-1$, $k = 1,\ldots,n-1$. The remaining inequalities are then fulfilled.

In [36], a matrix was called *equilibrated* if all its row sums and all its column sums are equal to zero. It was shown there that by appropriately subtracting constant rows and constant columns, every anti-Monge matrix can be transformed into an equilibrated anti-Monge matrix; interestingly enough, the product of such matrices (if they can be multiplied) is also an equilibrated anti-Monge matrix.

2

Solvability of systems of interval linear equations and inequalities

J. Rohn

2.1 Introduction and notations

This chapter deals with solvability and feasibility (i.e., nonnegative solvability) of systems of interval linear equations and inequalities. After a few preliminary sections, we delineate in Section 2.6 eight decision problems (weak solvability of equations through strong feasibility of inequalities) that are then solved in eight successive sections 2.7 to 2.14. It turns out that four problems are solvable in polynomial time and four are NP-hard. Some of the results are easy (Theorem 2.13), some difficult to prove (Theorem 2.14), and some are surprising (Theorem 2.24). Although solutions of several of them are already known, the complete classification of the eight problems given here is new. Some special cases (tolerance, control and algebraic solutions, systems with square matrices) are treated in Sections 2.16 to 2.19. The last, Section 2.21 contains additional notes and references to the material of this chapter. Some of the results find later applications in interval linear programming (Chapter 3).

We use the following notations. The ith row of a matrix A is denoted by $A_{i.}$ and the jth column by $A_{.j}$. For two matrices A, B of the same size, inequalities like $A \leq B$ or $A < B$ are understood componentwise. A is called nonnegative if $0 \leq A$; A^T is the transpose of A. The absolute value of a matrix $A = (a_{ij})$ is defined by $|A| = (|a_{ij}|)$. We use the following easy-to-prove properties valid whenever the respective operations and inequalities are defined.

(i) $A \leq B$ and $0 \leq C$ imply $AC \leq BC$.
(ii) $A \leq |A|$.
(iii) $|A| \leq B$ if and only if $-B \leq A \leq B$.
(iv) $|A + B| \leq |A| + |B|$.
(v) $||A| - |B|| \leq |A - B|$.
(vi) $|AB| \leq |A||B|$.

The same notations and results also apply to vectors that are always considered one-column matrices. Hence, for $a = (a_i)$ and $b = (b_i)$, $a^T b = \sum_i a_i b_i$

is the scalar product whereas ab^T is the matrix $(a_i b_j)$. Maximum (or minimum) of two vectors a, b is understood componentwise: i.e., $(\max\{a, b\})_i = \max\{a_i, b_i\}$ for each i. In particular, for vectors a^+, a^- defined by $a^+ = \max\{a, 0\}$, $a^- = \max\{-a, 0\}$ we have $a = a^+ - a^-$, $|a| = a^+ + a^-$, $a^+ \geq 0$, $a^- \geq 0$ and $(a^+)^T a^- = 0$. I denotes the unit matrix, e_j is the jth column of I and $e = (1, \ldots, 1)^T$ is the vector of all ones (in these cases we do not designate explicitly the dimension which can always be inferred from the context). In our descriptions to follow, an important role is played by the set Y_m of all ± 1 vectors in $\mathbb{R}^m$; i.e.,

$$Y_m = \{y \in \mathbb{R}^m \mid |y| = e\}.$$

Obviously, the cardinality of Y_m is 2^m. For each $x \in \mathbb{R}^m$ we define its sign vector $\operatorname{sgn} x$ by

$$(\operatorname{sgn} x)_i = \begin{cases} 1 \text{ if } x_i \geq 0, \\ -1 \text{ if } x_i < 0 \end{cases} \qquad (i = 1, \ldots, m),$$

so that $\operatorname{sgn} x \in Y_m$. For a given vector $y \in \mathbb{R}^m$ we denote

$$T_y = \operatorname{diag}(y_1, \ldots, y_m) = \begin{pmatrix} y_1 & 0 & \cdots & 0 \\ 0 & y_2 & \cdots & 0 \\ \vdots & \vdots & \ddots & \vdots \\ 0 & 0 & \cdots & y_m \end{pmatrix}. \tag{2.1}$$

With a few exceptions (mainly in the proof of Theorem 2.9), we use the notation T_y for vectors $y \in Y_m$ only, in which case we have $T_{-y} = -T_y$, $T_y^{-1} = T_y$ and $|T_y| = I$. For each $x \in \mathbb{R}^m$ we can write $|x| = T_z x$, where $z = \operatorname{sgn} x$; we often use this trick to remove the absolute value of a vector. Notice that $T_z x = (z_i x_i)_{i=1}^m$.

2.2 An algorithm for generating Y_m

It will prove helpful at a later stage to generate all the ± 1-vectors forming the set Y_m systematically one-by-one in such a way that any two successive vectors differ in exactly one entry. We describe here an algorithm for performing this task, formulated in terms of generating the whole set Y_m; in later applications the last-but-one line "$Y := Y \cup \{y\}$" is replaced by the respective action on the current vector y. The algorithm employs an auxiliary $(0, 1)$-vector $z \in \mathbb{R}^m$ used for determining the index k for which the current value of y_k should be changed to $-y_k$, and its description is as follows.

$z := 0 \in \mathbb{R}^m$; select $y \in Y_m$; $Y := \{y\}$;
while $z \neq e$
   $k := \min\{i \mid z_i = 0\}$;
   for $i := 1$ **to** $k-1$, $z_i := 0$; **end**
   $z_k := 1$; $y_k := -y_k$;
   $Y := Y \cup \{y\}$;
end
% $Y = Y_m$

Theorem 2.1. *For each $m \geq 1$ the algorithm at the output yields the set $Y = Y_m$ independently of the choice of the initial vector y.*

Proof. We prove the assertion by induction on m. For $m = 1$ it is a matter of simple computation to verify that the algorithm, if started from $y = 1$, generates $Y = \{1, -1\}$, and if started from $y = -1$, generates $Y = \{-1, 1\}$; in both cases $Y = Y_1$. Thus let the assertion hold for some $m - 1 \geq 1$ and let the algorithm be run for m. To see what is being done in the course of the algorithm, let us notice that in the main loop the initial string of the form

$$(1, 1, \ldots, 1, 0, \ldots)^T$$

of the current vector z is being found, where 0 is at the kth position, and it is being changed to

$$(0, 0, \ldots, 0, 1, \ldots)^T$$

until the vector z of all ones is reached (the last vector preceding it is $(0, 1, \ldots, 1, 1)^T$). Hence if we start the algorithm for m, then the sequence of vectors z and y, restricted to their first $m - 1$ entries, is the same as if the algorithm were run for $m - 1$, until vector z of the form

$$(1, 1, \ldots, 1, 0)^T \tag{2.2}$$

is reached. By that time, according to the induction hypothesis, the algorithm has constructed all the vectors $y \in Y_m$ with y_m being fixed throughout at its initial value. In the next step the vector (2.2) is switched to

$$(0, 0, \ldots, 0, 1)^T$$

and y_m is switched to $-y_m$. Now, from the point of view of the first $m - 1$ entries, the algorithm again starts from zero vector z and due to the induction hypothesis it again generates all the $(m - 1)$-dimensional ± 1-vectors in the first $m - 1$ entries, this time with the opposite value of y_m. This implies that at the end (when vector z of all ones is reached) the whole set Y_m is generated, which completes the proof by induction. □

We have needed a description starting from an arbitrary $y \in Y_m$ for the purposes of the proof by induction only; in practice we usually start with

$y = e$. The performance of the algorithm for $m = 3$ is illustrated in the following table. The algorithm is started from $z = 0$, $y = e$ (the first row) and the current values of z, y at the end of each pass through the "**while** ... **end**" loop are given in the next seven rows of the table.

z^T	y^T
(0, 0, 0)	(1, 1, 1)
(1, 0, 0)	(-1, 1, 1)
(0, 1, 0)	(-1, -1, 1)
(1, 1, 0)	(1, -1, 1)
(0, 0, 1)	(1, -1, -1)
(1, 0, 1)	(-1, -1, -1)
(0, 1, 1)	(-1, 1, -1)
(1, 1, 1)	(1, 1, -1)

2.3 Auxiliary complexity result

Given two vector norms $\|x\|_\alpha$ and $\|x\|_\beta$ in $\mathbb{R}^n$, a subordinate matrix norm $\|A\|_{\alpha,\beta}$ for $A \in \mathbb{R}^{n \times n}$ is defined by

$$\|A\|_{\alpha,\beta} = \max_{\|x\|_\alpha = 1} \|Ax\|_\beta \qquad (2.3)$$

(see Higham [51], p. 121). If we use the norms $\|x\|_1 = e^T|x| = \sum_i |x_i|$, $\|x\|_\infty = \max_i |x_i|$, then from (2.3) we obtain $\|A\|_{1,1} = \max_j \sum_i |a_{ij}|$, $\|A\|_{\infty,\infty} = \max_i \sum_j |a_{ij}|$, and $\|A\|_{1,\infty} = \max_{ij} |a_{ij}|$, so that all three norms are easy to compute. This, however, is no longer true for the fourth norm $\|A\|_{\infty,1}$. In [165] it is proved that

$$\|A\|_{\infty,1} = \max_{y \in Y_n} \|Ay\|_1 = \max_{z,y \in Y_n} z^T A y, \qquad (2.4)$$

where the set Y_n consists of 2^n vectors. One might hope to find an essentially better formula for $\|A\|_{\infty,1}$, but such an attempt is not likely to succeed due to the following complexity result proved again in [165].

Theorem 2.2. *The problem of checking whether*

$$\|A\|_{\infty,1} \geq 1$$

holds is NP-complete in the set of symmetric rational M-matrices.

A square matrix $A = (a_{ij})$ is called an M-matrix if $a_{ij} \leq 0$ for $i \neq j$ and $A^{-1} \geq 0$ (see p. 29). For our purposes it is advantageous to reformulate the result in terms of systems of inequalities.

Theorem 2.3. *The problem of checking whether a system of inequalities*

$$-e \leq Ax \leq e, \tag{2.5}$$

$$e^T |x| \geq 1 \tag{2.6}$$

has a solution is NP-complete in the set of nonnegative positive definite rational matrices.

Comment. Clearly, $e^T|x| = \|x\|_1$, so that the inequality (2.6) could be equivalently written as

$$\|x\|_1 \geq 1.$$

We prefer, however, the formulation given because terms of the form $e^T|x|$ arise quite naturally in the analysis of complexity of interval linear systems.

Proof. Given a symmetric rational M-matrix $A \in \mathbb{R}^{n \times n}$, consider the system

$$-e \leq A^{-1}x \leq e, \tag{2.7}$$

$$e^T |x| \geq 1 \tag{2.8}$$

which can be constructed in polynomial time since the same is true for A^{-1} (see Bareiss [8]). Since A is positive definite ([54], p. 114, assertion 2.5.3.3), A^{-1} is rational nonnegative positive definite. Obviously, the system (2.7), (2.8) has a solution if and only if

$$1 \leq \max\{e^T|x| \mid -e \leq A^{-1}x \leq e\} = \max\{e^T|Ax'| \mid -e \leq x' \leq e\}$$
$$= \max\{\|Ax'\|_1 \mid -e \leq x' \leq e\} = \max\{\|Ay\|_1 \mid y \in Y_n\} = \|A\|_{\infty,1}$$

holds, since the function $\|Ax'\|_1$ is convex over the unit cube $\{x' \mid -e \leq x' \leq e\}$ and therefore its maximum is attained at one of its vertices which are just the vectors in Y_n. Summing up, we have shown that $\|A\|_{\infty,1} \geq 1$ holds if and only if the system (2.7), (2.8) has a solution. Since the former problem is NP-complete (Theorem 2.2), the latter one is NP-hard; hence also the problem (2.5), (2.6) is NP-hard. Moreover, if (2.5), (2.6) has a solution, then, as we have seen, it also has a rational solution of the form $x = Ay$ for some $y \in Y_n$, and verification whether x solves (2.5), (2.6) can be performed in polynomial time. Hence the problem of checking solvability of (2.5), (2.6) belongs to the class NP and therefore it is NP-complete.   □

We later use this result to establish NP-hardness of several decision problems concerning systems of interval linear equations and inequalities. For a detailed introduction into complexity theory, see Garey and Johnson [41].

2.4 Solvability and feasibility

From this section on we consider systems of linear equations $Ax = b$ or systems of linear inequalities $Ax \leq b$. Unless said otherwise, it is always assumed that $A \in \mathbb{R}^{m \times n}$ and $b \in \mathbb{R}^m$, where m and n are arbitrary positive integers.

A system of linear equations $Ax = b$ is called *solvable* if it has a solution, and *feasible* if it has a nonnegative solution. Throughout this and the next chapter the reader is kindly asked to bear in mind that *feasibility means non-negative solvability*. The basic result concerning feasibility of linear equations was proved by Farkas [34] in 1902. As it is used at some crucial points in the sequel, we give here an elementary, but somewhat lengthy proof of it. The ideas of the proof are not exploited later, so that the reader may skip the proof without loss of continuity.

Theorem 2.4 (Farkas). *A system*

$$Ax = b \tag{2.9}$$

is feasible if and only if each p with $A^T p \geq 0$ satisfies $b^T p \geq 0$.

Proof. (a) If the system (2.9) has a solution $x \geq 0$ and if $A^T p \geq 0$ holds for some $p \in \mathbb{R}^m$, then $b^T p = (Ax)^T p = x^T (A^T p) \geq 0$. This proves the "only if" part of the theorem.

(b) We prove the "if" part by contradiction, proving that if the system (2.9) does not possess a nonnegative solution, then there exists a $p \in \mathbb{R}^m$ satisfying $A^T p \geq 0$ and $b^T p < 0$; for the purposes of the proof it is advantageous to write down this system in the column form

$$p^T A_{.j} \geq 0 \qquad (j = 1, \ldots, n), \tag{2.10}$$
$$p^T b < 0. \tag{2.11}$$

We prove this assertion by induction on n.

(b1) If $n = 1$, then A consists of a single column a. Let $W = \{\alpha a \mid \alpha \in \mathbb{R}\}$ be the subspace spanned by a. According to the orthogonal decomposition theorem (Meyer [88], p. 405), b can be written in the form

$$b = b_W + b_{W^\perp},$$

where $b_W \in W$ and $b_{W^\perp} \in W^\perp$, $W^\perp$ being the orthogonal complement of W. We consider two cases. If $b_{W^\perp} = 0$, then $b \in W$, so that $b = \alpha a$ for some $\alpha \in \mathbb{R}$. Since $Ax = b$ does not possess a nonnegative solution due to the assumption, it must be $\alpha < 0$ and $a \neq 0$, so that if we put $p = a$, then $p^T a = \|a\|_2^2 \geq 0$ and $p^T b = \alpha \|a\|_2^2 < 0$; hence p satisfies (2.10), (2.11). If $b_{W^\perp} \neq 0$, put $p = -b_{W^\perp}$; then $p^T a = 0$ and $p^T b = -\|b_{W^\perp}\|_2^2 < 0$, so that p again satisfies (2.10), (2.11).

(b2) Let the induction hypothesis hold for $n - 1 \geq 1$ and let a system (2.9), where $A \in \mathbb{R}^{m \times n}$, possess no nonnegative solution. Then neither does the system

$$\sum_{j=1}^{n-1} A_{\cdot j} x_j = b$$

(otherwise for $x_n = 0$ we would get a nonnegative solution of (2.9)); hence according to the induction hypothesis there exists a $\bar{p} \in \mathbb{R}^m$ satisfying

$$\bar{p}^T A_{\cdot j} \geq 0 \qquad (j = 1, \ldots, n-1), \tag{2.12}$$
$$\bar{p}^T b < 0. \tag{2.13}$$

If $\bar{p}^T A_{\cdot n} \geq 0$, then p satisfies (2.10), (2.11) and we are done. Thus assume that

$$\bar{p}^T A_{\cdot n} < 0. \tag{2.14}$$

Put

$$\alpha_j = \bar{p}^T A_{\cdot j} \qquad (j = 1, \ldots, n),$$
$$\beta = \bar{p}^T b;$$

then $\alpha_1 \geq 0, \ldots, \alpha_{n-1} \geq 0$, $\alpha_n < 0$ and $\beta < 0$. Consider the system

$$\sum_{j=1}^{n-1} (\alpha_n A_{\cdot j} - \alpha_j A_{\cdot n}) x_j = \alpha_n b - \beta A_{\cdot n}. \tag{2.15}$$

If it had a nonnegative solution $x_1, \ldots, x_{n-1}$, then we could rearrange it to the form

$$\sum_{j=1}^{n-1} A_{\cdot j} x_j + A_{\cdot n} x_n = b, \tag{2.16}$$

where

$$x_n = \frac{\beta - \sum_{j=1}^{n-1} \alpha_j x_j}{\alpha_n} > 0$$

due to (2.12), (2.13), (2.14), so that the system (2.16), and thus also (2.9), would have a nonnegative solution $x_1, \ldots, x_n$ contrary to the assumption. Therefore the system (2.15) does not possess a nonnegative solution and thus according to the induction hypothesis there exists a $\tilde{p}$ such that

$$\tilde{p}^T (\alpha_n A_{\cdot j} - \alpha_j A_{\cdot n}) \geq 0 \qquad (j = 1, \ldots, n-1), \tag{2.17}$$
$$\tilde{p}^T (\alpha_n b - \beta A_{\cdot n}) < 0. \tag{2.18}$$

Now we set

$$p = \alpha_n \tilde{p} - (\tilde{p}^T A_{\cdot n}) \bar{p}$$

and we show that p satisfies (2.10), (2.11). For $j = 1, \ldots, n-1$ we have according to (2.17),

$$p^T A_{\cdot j} = \alpha_n \tilde{p}^T A_{\cdot j} - (\tilde{p}^T A_{\cdot n}) \bar{p}^T A_{\cdot j} \geq \alpha_j \tilde{p}^T A_{\cdot n} - (\tilde{p}^T A_{\cdot n}) \alpha_j = 0; \tag{2.19}$$

for $j = n$ we get

$$p^T A_{.n} = \alpha_n \tilde{p}^T A_{.n} - (\tilde{p}^T A_{.n}) \bar{p}^T A_{.n} = \alpha_n \tilde{p}^T A_{.n} - (\tilde{p}^T A_{.n}) \alpha_n = 0, \qquad (2.20)$$

and finally from (2.18),

$$p^T b = \alpha_n \tilde{p}^T b - (\tilde{p}^T A_{.n}) \bar{p}^T b < \beta \tilde{p}^T A_{.n} - (\tilde{p}^T A_{.n}) \beta = 0, \qquad (2.21)$$

so that (2.19), (2.20), (2.21) imply (2.10) and (2.11); hence p is a vector having the asserted properties, which completes the proof by induction. □

With the help of the Farkas theorem we can now characterize solvability of systems of linear equations.

Theorem 2.5. *A system $Ax = b$ is solvable if and only if each p with $A^T p = 0$ satisfies $b^T p = 0$.*

Proof. If x solves $Ax = b$ and $A^T p = 0$ holds for some p, then $b^T p = p^T b = p^T Ax = (A^T p)^T x = 0$. Conversely, let the condition hold. Then for each p such that $A^T p \geq 0$ and $A^T p \leq 0$ we have $b^T p \geq 0$. But this, according to the Farkas theorem, is just the sufficient condition for the system

$$Ax_1 - Ax_2 = b \qquad (2.22)$$

to be feasible. Hence (2.22) has a solution $x_1 \geq 0$, $x_2 \geq 0$; thus $A(x_1 - x_2) = b$ and $Ax = b$ is solvable. □

For systems of linear inequalities we introduce the notions of solvability and feasibility in the same way: a system $Ax \leq b$ is called *solvable* if it has a solution, and *feasible* if it has a nonnegative solution. Again, we can use the Farkas theorem for characterizing solvability and feasibility.

Theorem 2.6. *A system $Ax \leq b$ is solvable if and only if each $p \geq 0$ with $A^T p = 0$ satisfies $b^T p \geq 0$.*

Proof. If x solves $Ax \leq b$ and $A^T p = 0$ holds for some $p \geq 0$, then $b^T p = p^T b \geq p^T Ax = 0$. Conversely, let the condition hold, so that each $p \geq 0$ with $A^T p \geq 0$, $A^T p \leq 0$ satisfies $b^T p \geq 0$. This, however, in view of the Farkas theorem means that the system

$$Ax_1 - Ax_2 + x_3 = b$$

is feasible. Hence due to the nonnegativity of x_3 we have $A(x_1 - x_2) \leq b$, and the system $Ax \leq b$ is solvable. □

Theorem 2.7. *A system $Ax \leq b$ is feasible if and only if each $p \geq 0$ with $A^T p \geq 0$ satisfies $b^T p \geq 0$.*

Proof. If $x \geq 0$ solves $Ax \leq b$ and $A^T p \geq 0$ holds for some $p \geq 0$, then $b^T p = p^T b = p^T Ax = (A^T p)^T x \geq 0$. Conversely, let the condition hold; then it is exactly the Farkas condition for the system

$$Ax_1 + x_2 = b \tag{2.23}$$

to be feasible. Hence (2.23) has a solution $x_1 \geq 0$, $x_2 \geq 0$, which implies $Ax_1 \leq b$, so that the system $Ax \leq b$ is feasible.   □

Finally, we sum up the results achieved in this section in the form of a table that reveals similarities and differences among the four necessary and sufficient conditions.

Problem	Condition
solvability of $Ax = b$	$(\forall p)(A^T p = 0 \Rightarrow b^T p = 0)$
feasibility of $Ax = b$	$(\forall p)(A^T p \geq 0 \Rightarrow b^T p \geq 0)$
solvability of $Ax \leq b$	$(\forall p \geq 0)(A^T p = 0 \Rightarrow b^T p \geq 0)$
feasibility of $Ax \leq b$	$(\forall p \geq 0)(A^T p \geq 0 \Rightarrow b^T p \geq 0)$

An important result published by Khachiyan [71] in 1979 says that feasibility of a system of linear equations can be checked (and a solution to it, if it exists, found) in polynomial time. Since all three other problems, as shown in the proofs, can be reduced to this one, it follows that all four problems can be solved in polynomial time.

2.5 Interval matrices and vectors

There are several ways to express inexactness of the data. One of them, which has particularly nice properties from the point of view of a user, employs the so-called interval matrices which we define in this section.

If $\underline{A}$, $\overline{A}$ are two matrices in $\mathbb{R}^{m \times n}$, $\underline{A} \leq \overline{A}$, then the set of matrices

$$\mathbf{A} = [\underline{A}, \overline{A}] = \{A \mid \underline{A} \leq A \leq \overline{A}\}$$

is called an interval matrix, and the matrices $\underline{A}$, $\overline{A}$ are called its bounds. Hence, if $\underline{A} = (\underline{a}_{ij})$ and $\overline{A} = (\overline{a}_{ij})$, then $\mathbf{A}$ is the set of all matrices $A = (a_{ij})$ satisfying

$$\underline{a}_{ij} \leq a_{ij} \leq \overline{a}_{ij} \tag{2.24}$$

for $i = 1, \ldots, m$, $j = 1, \ldots, n$. It is worth noting that each coefficient may attain any value in its interval (2.24) independently of the values taken on by other coefficients. Introducing additional relations among different coefficients makes interval problems much more difficult to solve and we do not follow this line in this chapter.

As shown later, in many cases it is more advantageous to express the data in terms of the center matrix

$$A_c = \tfrac{1}{2}(\underline{A} + \overline{A}) \tag{2.25}$$

and of the radius matrix

$$\Delta = \tfrac{1}{2}(\overline{A} - \underline{A}), \tag{2.26}$$

which is always nonnegative. From (2.25), (2.26) we easily obtain that

$$\underline{A} = A_c - \Delta,$$

$$\overline{A} = A_c + \Delta,$$

so that $\mathbf{A}$ can be given either as $[\underline{A}, \overline{A}]$, or as $[A_c - \Delta, A_c + \Delta]$, and consequently we can also write

$$\mathbf{A} = \{A \mid |A - A_c| \le \Delta\}.$$

In the sequel *we employ both forms and we switch freely between them* according to which one is more useful in the current context. The following proposition is the first example of usefulness of the center-radius notation.

Proposition 2.8. *Let* $\tilde{\mathbf{A}} = [\tilde{A}_c - \tilde{\Delta}, \tilde{A}_c + \tilde{\Delta}]$ *and* $\mathbf{A} = [A_c - \Delta, A_c + \Delta]$ *be interval matrices of the same size. Then* $\tilde{\mathbf{A}} \subseteq \mathbf{A}$ *if and only if*

$$|A_c - \tilde{A}_c| \le \Delta - \tilde{\Delta}$$

holds.

Proof. If $\tilde{\mathbf{A}} \subseteq \mathbf{A}$, then from

$$A_c - \Delta \le \tilde{A}_c - \tilde{\Delta} \le \tilde{A}_c + \tilde{\Delta} \le A_c + \Delta \tag{2.27}$$

we obtain

$$-(\Delta - \tilde{\Delta}) \le A_c - \tilde{A}_c \le \Delta - \tilde{\Delta}, \tag{2.28}$$

which gives

$$|A_c - \tilde{A}_c| \le \Delta - \tilde{\Delta}. \tag{2.29}$$

Conversely, (2.29) implies (2.28) and (2.27); hence $\tilde{\mathbf{A}} \subseteq \mathbf{A}$.   □

For an interval matrix $\mathbf{A} = [\underline{A}, \overline{A}] = [A_c - \Delta, A_c + \Delta]$, its transpose is defined by $\mathbf{A}^T = \{A^T \mid A \in \mathbf{A}\}$. Obviously, $\mathbf{A}^T = [\underline{A}^T, \overline{A}^T] = [A_c^T - \Delta^T, A_c^T + \Delta^T]$.

A special case of an interval matrix is an interval vector which is a one-column interval matrix

$$\mathbf{b} = \{b \mid \underline{b} \le b \le \overline{b}\},$$

where $\underline{b}, \overline{b} \in \mathbb{R}^m$. We again use the center vector

$$b_c = \tfrac{1}{2}(\underline{b} + \overline{b})$$

and the nonnegative radius vector

$$\delta = \tfrac{1}{2}(\overline{b} - \underline{b}),$$

and we employ both forms $\mathbf{b} = [\underline{b}, \overline{b}] = [b_c - \delta, b_c + \delta]$. Notice that interval matrices and vectors are typeset in boldface letters.

Given an $m \times n$ interval matrix $\mathbf{A} = [A_c - \Delta, A_c + \Delta]$, we define matrices

$$A_{yz} = A_c - T_y \Delta T_z \qquad (2.30)$$

for each $y \in Y_m$ and $z \in Y_n$ (T_y is given by (2.1)). The definition implies that

$$(A_{yz})_{ij} = (A_c)_{ij} - y_i \Delta_{ij} z_j = \begin{cases} \overline{a}_{ij} \text{ if } y_i z_j = -1, \\ \underline{a}_{ij} \text{ if } y_i z_j = 1 \end{cases}$$

($i = 1, \ldots, m$, $j = 1, \ldots, n$), so that $A_{yz} \in \mathbf{A}$ for each $y \in Y_m$, $z \in Y_n$. This finite set of matrices from $\mathbf{A}$ (of cardinality at most 2^{m+n-1} because $A_{yz} = A_{-y,-z}$ for each $y \in Y_m$, $z \in Y_n$), introduced in [156], plays an important role because it turns out that many problems with interval-valued data can be characterized in terms of these matrices, thereby obtaining finite characterizations of problems involving infinitely many sets of data. In the theorems to follow we show several examples of this approach, the most striking one being Theorem 2.14. We write A_{-yz} instead of $A_{-y,z}$. In particular, we have $A_{ye} = A_c - T_y \Delta$, $A_{ez} = A_c - \Delta T_z$, $A_{ee} = \underline{A}$ and $A_{-ee} = \overline{A}$.

For an m-dimensional interval vector $\mathbf{b} = [b_c - \delta, b_c + \delta]$, in analogy with matrices A_{yz} we define vectors

$$b_y = b_c + T_y \delta$$

for each $y \in Y_m$. Then for each such a y we have

$$(b_y)_i = (b_c)_i + y_i \delta_i = \begin{cases} \underline{b}_i \text{ if } y_i = -1, \\ \overline{b}_i \text{ if } y_i = 1 \end{cases}$$

($i = 1, \ldots, m$), so that $b_y \in \mathbf{b}$ for each $y \in Y_m$. In particular, $b_{-e} = \underline{b}$ and $b_e = \overline{b}$. Together with matrices A_{yz}, vectors b_y are used in finite characterizations of interval problems having right-hand sides.

2.6 Weak and strong solvability/feasibility

Let $\mathbf{A}$ be an $m \times n$ interval matrix and $\mathbf{b}$ an m-dimensional interval vector. Under a system of interval linear equations

$$\mathbf{A}x = \mathbf{b} \qquad (2.31)$$

we understand the *family* of all systems of linear equations

$$Ax = b \qquad (2.32)$$

with data satisfying
$$A \in \mathbf{A}, b \in \mathbf{b}, \tag{2.33}$$
and similarly a system of interval linear inequalities
$$\mathbf{A}x \le \mathbf{b} \tag{2.34}$$
is the *family* of all systems
$$Ax \le b$$
whose data satisfy
$$A \in \mathbf{A}, b \in \mathbf{b}.$$

We introduce the following definitions. A system (2.31) is said to be *weakly* solvable (feasible) if *some* system (2.32) with data (2.33) is solvable (feasible), and it is called *strongly* solvable (feasible) if *each* system (2.32) with data (2.33) is solvable (feasible). In the same way we define weak and strong solvability (feasibility) of a system of interval linear inequalities (2.34). Hence, the word "weakly" refers to validity of the respective property for some system in the family whereas the word "strongly" refers to its validity for all systems in the family.

Introduction of weak and strong properties has an obvious motivation. Assume we are to decide whether some system $A_0x = b_0$ is solvable, but the exact data of this system are not directly available to us (they come from some measurements, are afflicted with rounding errors, etc.); instead, we only know that they satisfy $A_0 \in \mathbf{A}$, $b_0 \in \mathbf{b}$. Then we can be sure that our system $A_0x = b_0$ is solvable only if we know that the system (2.31) is strongly solvable, and in a similar way we can be sure that the system $A_0x = b_0$ is not solvable only if we know that the system (2.31) is not weakly solvable. A similar reasoning also holds for feasibility and for interval linear inequalities.

In this way, combining weak and strong solvability or feasibility of systems of interval linear equations or inequalities, we arrive at eight decision problems:

- Weak solvability of equations,
- Weak feasibility of equations,
- Strong solvability of equations,
- Strong feasibility of equations,
- Weak solvability of inequalities,
- Weak feasibility of inequalities,
- Strong solvability of inequalities,
- Strong feasibility of inequalities.

We study these problems separately in the next eight sections. It is shown that all of them can be solved by finite means, however, in half of the cases the number of steps is exponential in matrix size and the respective problems are proved to be NP-hard.

2.7 Weak solvability of equations

In this section we study the first of the eight decision problems delineated in Section 2.6, namely weak solvability of systems of interval linear equations. As before, we assume that $\mathbf{A}$ is an $m \times n$ interval matrix and $\mathbf{b}$ is an m-dimensional interval vector, where m and n are arbitrary positive integers.

First we introduce a useful auxiliary term: a vector $x \in \mathbb{R}^n$ is called a *weak solution* of $\mathbf{A}x = \mathbf{b}$ if it satisfies $Ax = b$ for some $A \in \mathbf{A}$, $b \in \mathbf{b}$. Oettli and Prager [112] proved in 1964 the following nice and far-reaching characterization of weak solutions.

Theorem 2.9 (Oettli–Prager). *A vector $x \in \mathbb{R}^n$ is a weak solution of* $\mathbf{A}x = \mathbf{b}$ *if and only if it satisfies*

$$|A_c x - b_c| \leq \Delta|x| + \delta. \tag{2.35}$$

Proof. If x is a weak solution, then $Ax = b$ for some $A \in \mathbf{A}$, $b \in \mathbf{b}$, which gives $|A_c x - b_c| = |(A_c - A)x + b - b_c| \leq \Delta|x| + \delta$. Conversely, let $|A_c x - b_c| \leq \Delta|x| + \delta$ hold for some x. Define $y \in \mathbb{R}^m$ by

$$y_i = \begin{cases} \frac{(A_c x - b_c)_i}{(\Delta|x| + \delta)_i} & \text{if } (\Delta|x| + \delta)_i > 0, \\ 1 & \text{if } (\Delta|x| + \delta)_i = 0 \end{cases} \quad (i = 1, \ldots, m), \tag{2.36}$$

then $|y| \leq e$ and

$$A_c x - b_c = T_y(\Delta|x| + \delta). \tag{2.37}$$

Put $z = \operatorname{sgn} x$; then $|x| = T_z x$ and from (2.37) we get

$$(A_c - T_y \Delta T_z)x = b_c + T_y \delta. \tag{2.38}$$

Since $|y| \leq e$ and $z \in Y_n$, we have $|T_y \Delta T_z| \leq \Delta$ and $|T_y \delta| \leq \delta$, so that $A_c - T_y \Delta T_z \in \mathbf{A}$ and $b_c + T_y \delta \in \mathbf{b}$, which implies that x is a weak solution of $\mathbf{A}x = \mathbf{b}$. $\qquad\square$

The main merit of the Oettli–Prager theorem consists in the fact that it describes the set of all weak solutions by means of a single, but nonlinear, inequality (2.35). In the proof we have also established a constructive result that is worth stating independently.

Proposition 2.10. *If x solves (2.35), then it satisfies (2.38), where y is given by (2.36) and $z = \operatorname{sgn} x$.*

Weak solvability of a system $\mathbf{A}x = \mathbf{b}$, as it was defined in Section 2.6, is equivalent to existence of a weak solution to it. Hence we can employ the Oettli–Prager theorem to characterize weak solvability of interval linear equations. Let us recall that in accordance with the general definition (2.30) we have $A_{ez} = A_c - \Delta T_z$ and $A_{-ez} = A_c + \Delta T_z$.

Theorem 2.11. *A system* $\mathbf{A}x = \mathbf{b}$ *is weakly solvable if and only if the system*

$$A_{ez}x \le \overline{b}, \tag{2.39}$$
$$-A_{-ez}x \le -\underline{b} \tag{2.40}$$

is solvable for some $z \in Y_n$.

Proof. If $\mathbf{A}x = \mathbf{b}$ is weakly solvable, then it has a weak solution x that according to Theorem 2.9 satisfies (2.35) and thus also

$$-\Delta|x| - \delta \le A_c x - b_c \le \Delta|x| + \delta. \tag{2.41}$$

If we put $z = \operatorname{sgn} x$, then $|x| = T_z x$ and (2.41) turns into $A_{ez}x = (A_c - \Delta T_z)x \le b_c + \delta = \overline{b}$ and $A_{-ez}x = (A_c + \Delta T_z)x \ge b_c - \delta = \underline{b}$ which shows that x satisfies (2.39), (2.40). Conversely, let (2.39), (2.40) hold for some x and $z \in Y_n$. Then we have

$$-\Delta T_z x - \delta \le A_c x - b_c \le \Delta T_z x + \delta$$

and consequently

$$|A_c x - b_c| \le \Delta T_z x + \delta \le \Delta|x| + \delta;$$

hence x satisfies (2.35) and therefore it is a weak solution of $\mathbf{A}x = \mathbf{b}$. □

This result shows that checking weak solvability of interval linear equations can be in principle performed by finite means by checking solvability of systems (2.39), (2.40), $z \in Y_n$ by some finite procedure (e.g., a linear programming technique). However, to verify that $\mathbf{A}x = \mathbf{b}$ is not weakly solvable, we have to check all the systems (2.39), (2.40), $z \in Y_n$, whose number in the worst case is 2^n. Clearly, this is nearly impossible even for relatively small values of n (say, $n = 30$). It turns out that the source of these difficulties does not lie with inadequateness of our description, but that it is inherently present in the problem itself which is NP-hard. In the proof of this statement we show an approach that is also used several times later, namely a polynomial-time reduction of our standard NP-complete problem from Theorem 2.3 to the current problem, which proves its NP-hardness.

Theorem 2.12. *Checking weak solvability of interval linear equations is NP-hard.*

Proof. Let A be a square matrix. We first prove that the system

$$-e \le Ax \le e, \tag{2.42}$$

$$e^T|x| \ge 1 \tag{2.43}$$

has a solution if and only if the system of interval linear equations

$$[A, \overline{A}]x = [-e, e],$$ (2.44)

$$[-e^T, e^T]x = [1, 1]$$ (2.45)

is weakly solvable. If x solves (2.42), (2.43) and if we set $x' = \frac{x}{e^T|x|}$, then $|Ax'| = \frac{1}{e^T|x|}|Ax| \leq |Ax| \leq e$ and $e^T|x'| = 1$; hence x' satisfies $Ax' = b$, $z^T x' = 1$ for some $b \in [-e, e]$ and $z^T = (\operatorname{sgn} x')^T \in [-e^T, e^T]$, which means that (2.44), (2.45) is weakly solvable. Conversely, let (2.44), (2.45) have a weak solution x; then $Ax = b$ and $c^T x = 1$ for some $b \in [-e, e]$ and $c^T \in [-e^T, e^T]$; hence $|Ax| \leq e$ and $1 = c^T x \leq |c|^T |x| \leq e^T |x|$, so that x solves (2.42), (2.43). We have shown that the problem of checking solvability of (2.42), (2.43) can be reduced in polynomial time to that of checking weak solvability of (2.44), (2.45). Since the former problem is NP-complete by Theorem 2.3, the latter one is NP-hard. □

2.8 Weak feasibility of equations

Using the notion of a weak solution introduced in Section 2.7, we can say that a system $\mathbf{A}x = \mathbf{b}$ is weakly feasible (in the sense of the definition made in Section 2.6) if and only if it has a nonnegative weak solution. Hence we can again use the Oettli–Prager theorem to obtain a characterization of weak feasibility.

Theorem 2.13. *A system $\mathbf{A}x = \mathbf{b}$ is weakly feasible if and only if the system*

$$\underline{A}x \leq \overline{b},$$ (2.46)

$$-\overline{A}x \leq -\underline{b}$$ (2.47)

is feasible.

Proof. If $\mathbf{A}x = \mathbf{b}$ is weakly feasible, then it possesses a nonnegative weak solution x that by Theorem 2.9 satisfies

$$|A_c x - b_c| \leq \Delta x + \delta$$ (2.48)

and thus also

$$-\Delta x - \delta \leq A_c x - b_c \leq \Delta x + \delta,$$ (2.49)

which is (2.46), (2.47). Conversely, if (2.46), (2.47) has a nonnegative solution x, then it satisfies (2.49) and (2.48) and by the same Theorem 2.9 it is a nonnegative weak solution to $\mathbf{A}x = \mathbf{b}$ which means that this system is weakly feasible. □

Hence, only one system of linear inequalities (2.46), (2.47) is to be checked in this case. Referring to the last paragraph of Section 2.4, we can conclude that checking weak feasibility of interval linear equations can be performed in polynomial time whereas checking weak solvability, as we have seen in Theorem 2.12, is NP-hard.

2.9 Strong solvability of equations

By definition (Section 2.6), $\mathbf{A}x = \mathbf{b}$ is strongly solvable if each system $Ax = b$ with $A \in \mathbf{A}$, $b \in \mathbf{b}$ is solvable. If $\underline{A}_{ij} < \overline{A}_{ij}$ for some i, j or $\underline{b}_i < \overline{b}_i$ for some i, then the family $\mathbf{A}x = \mathbf{b}$ consists of infinitely many linear systems. Therefore the fact that solvability of these infinitely many systems can be characterized in terms of feasibility of finitely many systems is nontrivial, and so is the proof of the following theorem which also establishes a useful additional property. Conv X denotes the convex hull of X, i.e., the intersection of all convex subsets of $\mathbb{R}^n$ containing X.

Theorem 2.14. *A system $\mathbf{A}x = \mathbf{b}$ is strongly solvable if and only if for each $y \in Y_m$ the system*

$$A_{ye}x^1 - A_{-ye}x^2 = b_y, \tag{2.50}$$

$$x^1 \geq 0, \ x^2 \geq 0 \tag{2.51}$$

has a solution x_y^1, x_y^2. Moreover, if this is the case, then for each $A \in \mathbf{A}$, $b \in \mathbf{b}$ the system $Ax = b$ has a solution in the set

$$\mathrm{Conv}\{x_y^1 - x_y^2 \mid y \in Y_m\}.$$

Proof. "Only if": Let $\mathbf{A}x = \mathbf{b}$ be strongly solvable. Assume to the contrary that (2.50), (2.51) does not have a solution for some $y \in Y_m$. Then the Farkas theorem implies existence of a $p \in \mathbb{R}^m$ satisfying

$$(A_c - T_y\Delta)^T p \geq 0, \tag{2.52}$$

$$(A_c + T_y\Delta)^T p \leq 0, \tag{2.53}$$

$$b_y^T p < 0. \tag{2.54}$$

Now (2.52) and (2.53) together give

$$\Delta^T T_y p \leq A_c^T p \leq -\Delta^T T_y p;$$

hence

$$|A_c^T p| \leq -\Delta^T T_y p = |-\Delta^T T_y p| \leq \Delta^T |p|,$$

and the Oettli–Prager theorem as applied to the system $[A_c^T - \Delta^T, A_c^T + \Delta^T]x = [0, 0]$ shows that there exists a matrix $A \in \mathbf{A}$ such that

$$A^T p = 0. \tag{2.55}$$

In the light of Theorem 2.5, (2.55) and (2.54) mean that the system

$$Ax = b_y$$

has no solution, which contradicts our assumption of strong solvability since $A \in \mathbf{A}$ and $b_y \in \mathbf{b}$.

"If": Conversely, let for each $y \in Y_m$ the system (2.50), (2.51) have a solution x_y^1, x_y^2. Let $A \in \mathbf{A}$ and $b \in \mathbf{b}$. To prove that the system $Ax = b$ has a solution, take an arbitrary $y \in Y_m$ and put $x_y = x_y^1 - x_y^2$. Then we have

$$T_y(Ax_y - b) = T_y(A_c x_y - b_c) + T_y(A - A_c)x_y + T_y(b_c - b)$$
$$\geq T_y(A_c x_y - b_c) - \Delta|x_y| - \delta$$

since $|T_y(A - A_c)x_y| \leq \Delta|x_y|$, which implies $T_y(A - A_c)x_y \geq -\Delta|x_y|$, and similarly $|T_y(b_c - b)| \leq \delta$ implies $T_y(b_c - b) \geq -\delta$; thus

$$T_y(Ax_y - b) \geq T_y(A_c(x_y^1 - x_y^2) - b_c) - \Delta|x_y^1 - x_y^2| - \delta$$
$$\geq T_y(A_c(x_y^1 - x_y^2) - b_c) - \Delta(x_y^1 + x_y^2) - \delta$$
$$= T_y((A_c - T_y\Delta)x_y^1 - (A_c + T_y\Delta)x_y^2 - (b_c + T_y\delta))$$
$$= T_y(A_{ye}x_y^1 - A_{-ye}x_y^2 - b_y)$$
$$= 0$$

since x_y^1, x_y^2 solve (2.50), (2.51). In this way we have proved that for each $y \in Y_m$, x_y satisfies

$$T_y Ax_y \geq T_y b. \tag{2.56}$$

Using (2.56), we next prove that the system of linear equations

$$\sum_{y \in Y_m} \lambda_y Ax_y = b, \tag{2.57}$$

$$\sum_{y \in Y_m} \lambda_y = 1 \tag{2.58}$$

has a solution $\lambda_y \geq 0$, $y \in Y_m$. In view of the Farkas theorem, it suffices to show that for each $p \in \mathbb{R}^m$ and each $p_0 \in \mathbb{R}$,

$$p^T Ax_y + p_0 \geq 0 \text{ for each } y \in Y_m \tag{2.59}$$

implies

$$p^T b + p_0 \geq 0. \tag{2.60}$$

Thus let p and p_0 satisfy (2.59). Put $y = -\operatorname{sgn} p$; then $p = -T_y|p|$ and from (2.56), (2.59) we have

$$p^T b + p_0 = -|p|^T T_y b + p_0 \geq -|p|^T T_y Ax_y + p_0 = p^T Ax_y + p_0 \geq 0,$$

which proves (2.60). Hence the system (2.57), (2.58) has a solution $\lambda_y \geq 0$, $y \in Y_m$. Put $x = \sum_{y \in Y_m} \lambda_y x_y$; then $Ax = b$ by (2.57) and x belongs to the set $\operatorname{Conv}\{x_y \mid y \in Y_m\} = \operatorname{Conv}\{x_y^1 - x_y^2 \mid y \in Y_m\}$ by (2.58). This proves the "if" part, and also the additional assertion. $\square$

Let us have a closer look at the form of the systems (2.50). If $y_k = 1$, then the kth rows of A_{ye} and A_{-ye} are equal to the kth rows of $\underline{A}$ and $\overline{A}$, respectively, and $(b_y)_k = \overline{b}_k$. This means that in this case the kth equation of (2.50) has the form

$$(\underline{A}x^1 - \overline{A}x^2)_k = \overline{b}_k, \tag{2.61}$$

and similarly in case $y_k = -1$ it is of the form

$$(\overline{A}x^1 - \underline{A}x^2)_k = \underline{b}_k. \tag{2.62}$$

Hence we can see that the family of systems (2.50) for all $y \in Y_m$ is just the family of all systems whose kth equations are either of the form (2.61), or of the form (2.62) for $k = 1, \ldots, m$. Now we can use the algorithm of Section 2.2 to generate the systems $A_{ye}x^1 - A_{-ye}x^2 = b_y$ in such a way that any pair of successive systems differs in exactly one equation. In this way, a feasible solution x^1, x^2 of the preceding system satisfies all but at most one of the equations of the next generated system, so that this solution x^1, x^2 can be used as the initial iteration for the procedure for checking feasibility of the next system (the procedure is not specified in the algorithm; e.g., phase I of the simplex method may be used for this purpose). The complete description of the algorithm is as follows.

```
z := 0; y := e; strosolv := true;
A := A; B := A; b := b;
if Ax¹ − Bx² = b is not feasible then strosolv := false; end
while z ≠ e & strosolv
    k := min{i | zᵢ = 0};
    for i := 1 to k − 1, zᵢ := 0; end
    zₖ := 1; yₖ := −yₖ;
    if yₖ = 1 then Aₖ. := Aₖ.; Bₖ. := Aₖ.; bₖ := bₖ;
              else Aₖ. := Aₖ.; Bₖ. := Aₖ.; bₖ := bₖ;
    end
    if Ax¹ − Bx² = b is not feasible then strosolv := false; end
end
% Ax = b is strongly solvable if and only if strosolv = true.
```

A small change can greatly improve the performance of the algorithm. Observe that if

$$\underline{A}_{k.} = \overline{A}_{k.} \quad \text{and} \quad \underline{b}_k = \overline{b}_k \tag{2.63}$$

hold for some k, then the equations (2.61) and (2.62) are the same and there is no need to solve the same system anew. Hence only rows satisfying

$$\underline{A}_{k.} \neq \overline{A}_{k.} \quad \text{or} \quad \underline{b}_k < \overline{b}_k \tag{2.64}$$

play any role. Let us reorder the equations of $\mathbf{A}x = \mathbf{b}$ so that those satisfying (2.64) go first, followed by those with (2.63). Hence, for the reordered system the matrix (Δ, δ) has first q nonzero rows, followed by $m - q$ zero rows $(0 \leq q \leq m)$. Now we can employ the algorithm in literally the same formulation, but started with $z := 0 \in \mathbb{R}^q$, $y := e \in \mathbb{R}^q$ (instead of $z, y \in \mathbb{R}^m$ in the original version). In this way, in the case of strong solvability 2^q systems $A_{ye}x^1 - A_{-ye}x^2 = b_y$ are to be checked for feasibility. Clearly, the whole procedure can be considered acceptable for moderate values of q only.

Since the number of systems to be checked is in the worst case exponential in the matrix size, we may suspect the problem to be NP-hard. It turns out to be indeed the case, and the NP-complete problem of Theorem 2.3 can again be used for the purpose of the proof of this result.

Theorem 2.15. *Checking strong solvability of interval linear equations is NP-hard.*

Proof. Let A be square $n \times n$. We prove that the system

$$-e \leq Ax \leq e, \tag{2.65}$$

$$e^T |x| \geq 1 \tag{2.66}$$

has a solution if and only if the system of interval linear equations

$$[A - ee^T, A + ee^T]x = [0, e] \tag{2.67}$$

is *not* strongly solvable. "If": Assume that (2.67) is not strongly solvable, so that $A'x = b'$ does not have a solution for some $A' \in [A - ee^T, A + ee^T]$ and $b' \in [0, e]$. Then A' must be singular; hence $A'x' = 0$ for some $x' \neq 0$. Then x' is a weak solution of the system $[A - ee^T, A + ee^T]x = [0, 0]$; hence $|Ax'| \leq ee^T |x'|$ by the Oettli–Prager theorem. Now if we set $x = \frac{x'}{e^T|x'|}$, then $|Ax| \leq e$ and $e^T |x| = 1$, so that x solves (2.65), (2.66). "Only if" by contradiction: Assume that (2.67) is strongly solvable, and let A' be an arbitrary matrix in $[A - ee^T, A + ee^T]$. Then for each $j = 1, \ldots, n$ the system $A'x = e_j$ (where $e_j \in [0, e]$ is the jth column of the unit matrix I) has a solution x^j; hence the matrix X consisting of columns $x^1, \ldots, x^n$ satisfies $A'X = I$, so that A' is nonsingular. Hence, strong solvability of (2.67) implies nonsingularity of each $A' \in [A - ee^T, A + ee^T]$. Assume now that (2.65), (2.66) has a solution x. Then $|Ax| \leq e \leq ee^T |x|$, and the Oettli–Prager theorem implies that x solves $A'x = 0$ for some $A' \in [A - ee^T, A + ee^T]$; hence A' is singular which contradicts the above fact that each $A' \in [A - ee^T, A + ee^T]$ is nonsingular. This contradiction shows that strong solvability of (2.67) precludes existence of a solution to (2.65), (2.66), which proves the "only if" part of the assertion. In view of the established equivalence, we can see that the problem of checking solvability of (2.65), (2.66) can be reduced in polynomial time to that of checking strong solvability of (2.67). By Theorem 2.3, the former problem is NP-complete; hence the latter one is NP-hard. $\square$

In an analogy with weak solutions, we may also introduce strong solutions of systems of interval linear equations. A vector x is said to be a *strong solution* of $\mathbf{A}x = \mathbf{b}$ if it satisfies $Ax = b$ for each $A \in \mathbf{A}$, $b \in \mathbf{b}$. We have this characterization of strong solutions:

Theorem 2.16. *A vector $x \in \mathbb{R}^n$ is a strong solution of $\mathbf{A}x = \mathbf{b}$ if and only if it satisfies*

$$A_c x = b_c, \tag{2.68}$$

$$\Delta|x| = \delta = 0. \tag{2.69}$$

Proof. Let x be a strong solution of $\mathbf{A}x = \mathbf{b}$. Put $z = \operatorname{sgn} x$; then $|x| = T_z x$, and x satisfies both

$$A_c x = b_c \tag{2.70}$$

and

$$(A_c + \Delta T_z)x = b_c - \delta. \tag{2.71}$$

Subtracting (2.70) from (2.71), we obtain

$$\Delta|x| = \Delta T_z x = -\delta,$$

where $\Delta|x| \geq 0$ and $-\delta \leq 0$; hence $\Delta|x| = \delta = 0$. Conversely, if (2.68) and (2.69) hold, then for each $A \in \mathbf{A}$, $b \in \mathbf{b}$ we have

$$|Ax - b| = |A_c x - b_c + (A - A_c)x + b_c - b| \leq \Delta|x| + \delta = 0,$$

so that $Ax = b$; hence x is a strong solution of $\mathbf{A}x = \mathbf{b}$. $\qquad\square$

The condition $\Delta|x| = 0$ in (2.69) says that it must be $x_j = 0$ for each j with $\Delta_{.j} \neq 0$. Hence, putting $J = \{j \mid \Delta_{.j} \neq 0\}$, we may reformulate (2.68), (2.69) in the form

$$\sum_{j \notin J} (A_c)_{.j} x_j = b_c, \tag{2.72}$$

$$x_j = 0 \qquad (j \in J), \tag{2.73}$$

$$\delta = 0, \tag{2.74}$$

which shows that checking existence of a strong solution (and, in the positive case, also computation of it) may be performed by solving a single system of linear equations (2.72). But on the whole the system (2.72)–(2.74) shows that strong solutions exist on rare occasions only, as could have been expected already from the definition.

2.10 Strong feasibility of equations

By definition in Section 2.6, a system $\mathbf{A}x = \mathbf{b}$ is strongly feasible if each system $Ax = b$ with $A \in \mathbf{A}$, $b \in \mathbf{b}$ is feasible. It turns out that characterization of strong feasibility can be easily derived from that of strong solvability.

Theorem 2.17. *A system $\mathbf{A}x = \mathbf{b}$ is strongly feasible if and only if for each $y \in Y_m$ the system*

$$A_{ye}x = b_y \tag{2.75}$$

has a nonnegative solution x_y. Moreover, if this is the case, then for each $A \in \mathbf{A}$, $b \in \mathbf{b}$ the system $Ax = b$ has a solution in the set

$$\mathrm{Conv}\{x_y \mid y \in Y_m\}.$$

Proof. If $\mathbf{A}x = \mathbf{b}$ is strongly feasible, then each system (2.75) has a nonnegative solution since $A_{ye} \in \mathbf{A}$ and $b_y \in \mathbf{b}$ for each $y \in Y_m$. Conversely, if for each $y \in Y_m$ the system (2.75) has a nonnegative solution x_y, then setting $x_y^1 = x_y$, $x_y^2 = 0$ for each $y \in Y_m$, we can see that x_y^1, x_y^2 solve (2.50), (2.51). This according to Theorem 2.14 means that each system $Ax = b$, $A \in \mathbf{A}$, $b \in \mathbf{b}$ has a solution in the set $\mathrm{Conv}\{x_y^1 - x_y^2 \mid y \in Y_m\} = \mathrm{Conv}\{x_y \mid y \in Y_m\}$ which is a part of the nonnegative orthant; hence $\mathbf{A}x = \mathbf{b}$ is strongly feasible. $\square$

Repeating the argument following the proof of Theorem 2.14, we can say that the kth row of (2.75) is of the form

$$(\underline{A}x)_k = \overline{b}_k$$

if $y_k = 1$ and of the form

$$(\overline{A}x)_k = \underline{b}_k$$

if $y_k = -1$. Hence, the algorithm for checking strong solvability can be easily adapted for the present purpose.

```
z := 0; y := e; strofeas := true;
A := A; b := b;
if Ax = b is not feasible then strofeas := false; end
while z ≠ e & strofeas
    k := min{i | z_i = 0};
    for i := 1 to k - 1, z_i := 0; end
    z_k := 1; y_k := -y_k;
    if y_k = 1 then A_k. := A_k.; b_k := b_k; else A_k. := A_k.; b_k := b_k; end
    if Ax = b is not feasible then strofeas := false; end
end
% Ax = b is strongly feasible if and only if strofeas = true.
```

As in Section 2.9, the equations of $\mathbf{A}x = \mathbf{b}$ should be first reordered so that the first q of them satisfy (2.64) and the last $m - q$ of them are of the form (2.63). Then the algorithm remains in force if it is initialized with $z := 0 \in \mathbb{R}^q$, $y := e \in \mathbb{R}^q$.

In contrast to checking weak feasibility which is polynomial-time (Section 2.8), checking strong feasibility remains NP-hard. The proof, going along similar lines as before, is a little bit different since $n \times 2n$ matrices are needed here.

Theorem 2.18. *Checking strong feasibility of interval linear equations is NP-hard.*

Proof. Let A be square $n \times n$. We prove that the system

$$-e \leq Ax \leq e, \tag{2.76}$$

$$e^T |x| \geq 1 \tag{2.77}$$

has a solution if and only if the system of interval linear equations

$$[(A^T - ee^T, -A^T - ee^T), (A^T + ee^T, -A^T + ee^T)]x = [-e, e] \tag{2.78}$$

(with an $n \times 2n$ interval matrix) is *not* strongly feasible. "If": Let (2.78) be not strongly feasible; then according to Theorem 2.17 there exists a $y \in Y_m$ such that the system $A_{ye}x = b_y$ is not feasible. In our case this system has the form

$$(A^T - ye^T)x^1 + (-A^T - ye^T)x^2 = y.$$

Since it is not feasible, the Farkas theorem assures existence of a vector x' satisfying

$$(A - ey^T)x' \geq 0, \tag{2.79}$$
$$(-A - ey^T)x' \geq 0, \tag{2.80}$$
$$y^T x' < 0; \tag{2.81}$$

then (2.79), (2.80) imply

$$|Ax'| \leq -ey^T x' = |-ey^T x'| \leq ee^T |x'|,$$

where $x' \neq 0$ by (2.81), hence the vector $x = \frac{x'}{e^T |x'|}$ satisfies $|Ax| \leq e$ and $e^T |x| = 1$, so that it is a solution to (2.76), (2.77). "Only if" by contradiction: Assume that (2.78) is strongly feasible. Let $A' \in [A - ee^T, A + ee^T]$; then $A'^T \in [A^T - ee^T, A^T + ee^T]$ and $-A'^T \in [-A^T - ee^T, -A^T + ee^T]$, so that strong feasibility of (2.78) implies that for each $j = 1, \ldots, n$ the equation

$$A'^T x^1 - A'^T x^2 = e_j$$

is feasible; i.e., the equation $A'^T x = e_j$ has a solution x^j. Then the matrix X consisting of columns $x^1, \ldots, x^n$ satisfies $A'^T X = I$, which proves that A'^T, and thus also A', is nonsingular. We have proved that strong feasibility of (2.78) implies nonsingularity of each $A' \in [A - ee^T, A + ee^T]$. As we have seen in the proof of Theorem 2.15, solvability of (2.76), (2.77) would mean existence of a singular matrix $A' \in [A - ee^T, A + ee^T]$, a contradiction. Hence (2.76), (2.77) is not solvable, which concludes the proof of the "only if" part. In view of Theorem 2.3, the established equivalence shows that checking strong feasibility is NP-hard.   □

Of the four decision problems related to interval linear equations we have investigated so far, three were found to be NP-hard and only one to be solvable in polynomial time. In the next four sections we show that this ratio becomes exactly reciprocal for interval linear inequalities: only one problem is NP-hard, and three are solvable in polynomial time.

2.11 Weak solvability of inequalities

As in Section 2.7, we first define $x \in \mathbb{R}^n$ to be a *weak solution* of a system of interval linear inequalities $\mathbf{A}x \leq \mathbf{b}$ if it satisfies $Ax \leq b$ for some $A \in \mathbf{A}$, $b \in \mathbf{b}$. Gerlach [43] proved in 1981 an analogue of the Oettli–Prager theorem for the case of interval linear inequalities.

Theorem 2.19 (Gerlach). *A vector x is a weak solution of $\mathbf{A}x \leq \mathbf{b}$ if and only if it satisfies*

$$A_c x - \Delta|x| \leq \overline{b}. \tag{2.82}$$

Proof. If x solves $Ax \leq b$ for some $A \in \mathbf{A}$ and $b \in \mathbf{b}$, then

$$A_c x - b_c \leq (A_c - A)x + b - b_c \leq |(A_c - A)x + b - b_c| \leq \Delta|x| + \delta,$$

which is (2.82). Conversely, let (2.82) hold for some x. Put $z = \operatorname{sgn} x$, then substituting $|x| = T_z x$ into (2.82) leads to

$$A_{ez} x \leq \overline{b},$$

where $A_{ez} \in \mathbf{A}$ and $\overline{b} \in \mathbf{b}$; hence x is a weak solution of $\mathbf{A}x \leq \mathbf{b}$.   □

A system $\mathbf{A}x \leq \mathbf{b}$ is weakly solvable (Section 2.6) if some system $Ax \leq b$, $A \in \mathbf{A}$, $b \in \mathbf{b}$ is solvable; in other words, weak solvability is equivalent to existence of a weak solution. Hence, Gerlach's theorem provides us with the following characterization.

Theorem 2.20. *A system $\mathbf{A}x \leq \mathbf{b}$ is weakly solvable if and only if the system*

$$A_{ez} x \leq \overline{b} \tag{2.83}$$

is solvable for some $z \in Y_n$.

Proof. If x is a weak solution of $\mathbf{A}x \leq \mathbf{b}$, then, as we have seen in the proof of the Gerlach theorem, it satisfies (2.83) for $z = \operatorname{sgn} x$. Conversely, if x satisfies (2.83) for some $z \in Y_n$, then it is a weak solution of the system $\mathbf{A}x \leq \mathbf{b}$ which is then weakly solvable. $\qquad\square$

The description suggests that the problem might be NP-hard, and it turns out to be again the case.

Theorem 2.21. *Checking weak solvability of interval linear inequalities is NP-hard.*

Proof. Given a square matrix A, the system

$$-e \leq Ax \leq e, \qquad (2.84)$$

$$e^T |x| \geq 1 \qquad (2.85)$$

can be rewritten equivalently as

$$\begin{pmatrix} A \\ -A \\ 0^T \end{pmatrix} x - \begin{pmatrix} 0 \\ 0 \\ e^T \end{pmatrix} |x| \leq \begin{pmatrix} e \\ e \\ -1 \end{pmatrix},$$

which is just the Gerlach inequality (2.82) for the system

$$\mathbf{A}x \leq \mathbf{b}, \qquad (2.86)$$

where

$$A_c = \begin{pmatrix} A \\ -A \\ 0^T \end{pmatrix}, \quad \Delta = \begin{pmatrix} 0 \\ 0 \\ e^T \end{pmatrix}, \quad \underline{b} = \overline{b} = \begin{pmatrix} e \\ e \\ -1 \end{pmatrix}. \qquad (2.87)$$

Hence the system (2.84), (2.85) has a solution if and only if the system of interval linear inequalities (2.86), (2.87) is weakly solvable. Thus the NP-complete problem of checking solvability of (2.84), (2.85) (Theorem 2.3) can be reduced in polynomial time to the problem of checking weak solvability of interval linear inequalities, which is then NP-hard. $\qquad\square$

2.12 Weak feasibility of inequalities

Weak feasibility of inequalities was defined in Section 2.6 as existence of a nonnegative weak solution. For nonnegative x we can replace the term $|x|$ in the Gerlach inequality simply by x, thereby obtaining this characterization:

Theorem 2.22. *A system $\mathbf{A}x \leq \mathbf{b}$ is weakly feasible if and only if the system*

$$\underline{A}x \leq \overline{b} \qquad (2.88)$$

is feasible.

Proof. If $x \geq 0$ satisfies $Ax \leq b$ for some $A \in \mathbf{A}$ and $b \in \mathbf{b}$, then

$$\underline{A}x \leq Ax \leq b \leq \overline{b}$$

and x is a feasible solution to (2.88). Conversely, feasibility of (2.88) obviously implies weak feasibility of $\mathbf{A}x \leq \mathbf{b}$.   □

Since feasibility of only one system of linear inequalities is to be checked, the problem is solvable in polynomial time (see the last paragraph of Section 2.4).

2.13 Strong solvability of inequalities

By definition, a system $\mathbf{A}x \leq \mathbf{b}$ is strongly solvable if each system $Ax \leq b$ with $A \in \mathbf{A}$, $b \in \mathbf{b}$ is solvable. Since the problem of checking strong solvability of interval linear equations is NP-hard (Theorem 2.15), one might expect the same to be the case for interval linear inequalities. But this analogy is no longer true, and we have this rather surprising result:

Theorem 2.23. *A system* $\mathbf{A}x \leq \mathbf{b}$ *is strongly solvable if and only if the system*

$$\overline{A}x^1 - \underline{A}x^2 \leq \underline{b} \tag{2.89}$$

is feasible.

Proof. "Only if": Assume to the contrary that the system (2.89) is not feasible; then neither is the system

$$\overline{A}x^1 - \underline{A}x^2 + x^3 = \underline{b},$$

and the Farkas theorem implies existence of a vector $p \in \mathbb{R}^m$ satisfying

$$\overline{A}^T p \geq 0, \tag{2.90}$$
$$\underline{A}^T p \leq 0, \tag{2.91}$$
$$p \geq 0, \tag{2.92}$$
$$\underline{b}^T p < 0. \tag{2.93}$$

Then (2.90) and (2.91) give

$$-\Delta^T p \leq -A_c^T p \leq \Delta^T p;$$

hence

$$|A_c^T p| \leq \Delta^T p = \Delta^T |p|$$

because of (2.92), and the Oettli–Prager theorem as applied to the system

$$[A_c^T - \Delta^T, A_c^T + \Delta^T]x = [0, 0]$$

implies existence of a matrix $A \in \mathbf{A}$ satisfying

$$A^T p = 0,$$

which together with (2.92) and (2.93) shows in the light of Theorem 2.6 that the system

$$Ax \leq \underline{b}$$

does not have a solution, a contradiction.

"If": Let $x^1 \geq 0$, $x^2 \geq 0$ solve (2.89). Then for each $A \in \mathbf{A}$ and each $b \in \mathbf{b}$ we have

$$A(x^1 - x^2) \leq \overline{A}x^1 - \underline{A}x^2 \leq \underline{b} \leq b,$$

so that $x^1 - x^2$ solves $Ax \leq b$. Hence $\mathbf{A}x \leq \mathbf{b}$ is strongly solvable; even more, all the systems $Ax \leq b$, $A \in \mathbf{A}$, $b \in \mathbf{b}$ share a common solution $x^1 - x^2$.   □

Hence checking strong solvability of inequalities can be performed in polynomial time. Let us call a vector x satisfying $Ax \leq b$ for each $A \in \mathbf{A}$, $b \in \mathbf{b}$ a *strong solution* of $\mathbf{A}x \leq \mathbf{b}$. We have simultaneously proved the following result.

Theorem 2.24. *If a system $\mathbf{A}x \leq \mathbf{b}$ is strongly solvable, then it has a strong solution.*

In other words, if each system $Ax \leq b$ with data satisfying $A \in \mathbf{A}$, $b \in \mathbf{b}$ has a solution of its own (depending on A and b, say $x(A, b)$), then all these systems share a common solution. This fact is certainly not obvious.

We have this characterization of strong solutions:

Theorem 2.25. *The following assertions are equivalent.*

(i) x is a strong solution of $\mathbf{A}x \leq \mathbf{b}$.
(ii) x satisfies

$$A_c x - b_c \leq -\Delta|x| - \delta. \tag{2.94}$$

(iii) $x = x^1 - x^2$, where x^1, x^2 satisfy

$$\overline{A}x^1 - \underline{A}x^2 \leq \underline{b}, \tag{2.95}$$

$$x^1 \geq 0, \ x^2 \geq 0. \tag{2.96}$$

Proof. We prove (i)$\Rightarrow$(ii)$\Rightarrow$(iii)$\Rightarrow$(i).

(i)$\Rightarrow$(ii): If $Ax \leq b$ for each $A \in \mathbf{A}$, $b \in \mathbf{b}$, then also $A_{-ez}x \leq \underline{b}$, where $z = \operatorname{sgn} x$; hence

$$A_c x + \Delta|x| = (A_c + \Delta T_z)x = A_{-ez}x \leq \underline{b} = b_c - \delta,$$

which implies (2.94).

(ii)$\Rightarrow$(iii): If x satisfies (2.94), then for $x^1 = x^+ = \max\{x, 0\}$, $x^2 = x^- = \max\{-x, 0\}$ we have $x^1 \geq 0$, $x^2 \geq 0$ and

$$\overline{A}x^1 - \underline{A}x^2 = A_c(x^1 - x^2) + \Delta(x^1 + x^2) = A_c x + \Delta|x| \le b_c - \delta = \underline{b};$$

hence x^1, x^2 solve (2.95), (2.96) and $x = x^1 - x^2$.

(iii)$\Rightarrow$(i) was proved in the "if" part of the proof of Theorem 2.23.   □

We can sum up these results in the form of a simple algorithm:

---

if (2.95), (2.96) has a solution x^1, x^2
then set $x := x^1 - x^2$ and terminate:
    x is a strong solution of $\mathbf{A}x \le \mathbf{b}$;
else terminate: $\mathbf{A}x \le \mathbf{b}$ is not strongly solvable;
end

2.14 Strong feasibility of inequalities

Finally, checking strong feasibility of inequalities is easy to characterize and can be done in polynomial time.

Theorem 2.26. *A system* $\mathbf{A}x \le \mathbf{b}$ *is strongly feasible if and only if the system*

$$\overline{A}x \le \underline{b} \tag{2.97}$$

is feasible.

Proof. If $\mathbf{A}x \le \mathbf{b}$ is strongly feasible, then (2.97) is feasible. Conversely, if (2.97) has a solution $x \ge 0$, then for each $A \in \mathbf{A}$, $b \in \mathbf{b}$ we have

$$Ax \le \overline{A}x \le \underline{b} \le b;$$

hence $\mathbf{A}x \le \mathbf{b}$ is strongly feasible.   □

2.15 Summary I: Complexity results

We can now summarize the results of the previous eight sections in the form of a table.

system of	equations	weak-ly	solvable	NP-hard
			feasible	polynomial-time
		strong-ly	solvable	NP-hard
			feasible	NP-hard
	inequa-lities	weak-ly	solvable	NP-hard
			feasible	polynomial-time
		strong-ly	solvable	polynomial-time
			feasible	polynomial-time

We can draw several conclusions from it. For interval problems, on the average:

(i)   Properties of equations are more difficult to check than those of inequalities;

(ii) Checking solvability is more difficult than checking feasibility; and

(iii) There is no such distinction between weak and strong properties.

2.16 Tolerance solutions

So far we have investigated mainly decision problems and in that frame four types of solutions (weak and strong solutions of both equations and inequalities) were introduced as auxiliary tools only. In this and in the next two sections we define three additional types of solutions motivated by some practical considerations.

In the present section we study tolerance solutions. A vector $x \in \mathbb{R}^n$ is said to be a *tolerance solution* of $\mathbf{A}x = \mathbf{b}$ if it satisfies $Ax \in \mathbf{b}$ for each $A \in \mathbf{A}$. The name of this type of solution reflects the fact that vector Ax stays within the prescribed tolerance $[\underline{b}, \overline{b}]$ independently of the choice of $A \in \mathbf{A}$. Original motivations for introducing and studying tolerance solutions came from the problem of crane construction (Nuding and Wilhelm [110]) and from the problem of input–output planning with inexact data [146].

The definition can also be recast by saying that x shall satisfy

$$\{Ax \mid A \in \mathbf{A}\} \subseteq \mathbf{b}. \tag{2.98}$$

We start therefore with a description of the left-hand-side set in (2.98).

Proposition 2.27. *Let $\mathbf{A}$ be an $m \times n$ interval matrix and let $x \in \mathbb{R}^n$. Then there holds*

$$\{Ax \mid A \in \mathbf{A}\} = [A_c x - \Delta|x|, A_c x + \Delta|x|]. \tag{2.99}$$

Proof. If $b \in \{Ax \mid A \in \mathbf{A}\}$, then $Ax = b$ for some $A \in \mathbf{A}$; hence x is a weak solution of

$$\mathbf{A}x = [b, b] \tag{2.100}$$

and by the Oettli–Prager theorem it satisfies

$$|A_c x - b| \leq \Delta|x|; \tag{2.101}$$

hence

$$-\Delta|x| \leq A_c x - b \leq \Delta|x| \tag{2.102}$$

and

$$A_c x - \Delta|x| \leq b \leq A_c x + \Delta|x|. \tag{2.103}$$

We have proved that $\{Ax \mid A \in \mathbf{A}\} \subseteq [A_c x - \Delta|x|, A_c x + \Delta|x|]$. Conversely, if $b \in [A_c x - \Delta|x|, A_c x + \Delta|x|]$, then b satisfies (2.103), (2.102), and (2.101); hence x is a weak solution of (2.100) which gives that $b \in \{Ax \mid A \in \mathbf{A}\}$. This proves the converse inclusion; hence (2.99) holds. □

With the help of this auxiliary result we can give two equivalent descriptions of tolerance solutions.

Theorem 2.28. *The following assertions are equivalent.*

(i) x is a tolerance solution of $\mathbf{A}x = \mathbf{b}$.

(ii) x satisfies

$$|A_c x - b_c| \leq -\Delta|x| + \delta. \qquad (2.104)$$

(iii) $x = x_1 - x_2$, where x_1, x_2 satisfy

$$\overline{A}x_1 - \underline{A}x_2 \leq \overline{b}, \qquad (2.105)$$

$$\underline{A}x_1 - \overline{A}x_2 \geq \underline{b}, \qquad (2.106)$$

$$x_1 \geq 0, \, x_2 \geq 0. \qquad (2.107)$$

Proof. We prove (i)$\Rightarrow$(ii)$\Rightarrow$(iii)$\Rightarrow$(i).

(i)$\Rightarrow$(ii): According to Proposition 2.27,

$$\{Ax \mid A \in \mathbf{A}\} = [A_c x - \Delta|x|, A_c x + \Delta|x|].$$

Hence, if x is a tolerance solution, then

$$[A_c x - \Delta|x|, A_c x + \Delta|x|] \subseteq [b_c - \delta, b_c + \delta],$$

which implies

$$b_c - \delta \leq A_c x - \Delta|x| \leq A_c x + \Delta|x| \leq b_c + \delta$$

and thus also

$$-(-\Delta|x| + \delta) \leq A_c x - b_c \leq -\Delta|x| + \delta, \qquad (2.108)$$

which is (2.104).

(ii)$\Rightarrow$(iii): If x satisfies (2.104), then for $x_1 = x^+$, $x_2 = x^-$ we have $x = x_1 - x_2$, $|x| = x_1 + x_2$ and the inequalities (2.108) turn into

$$\Delta(x_1 + x_2) - \delta \leq A_c(x_1 - x_2) - b_c \leq -\Delta(x_1 + x_2) + \delta,$$

which gives (2.105), (2.106), and (2.107) is satisfied because $x^+ \geq 0$, $x^- \geq 0$.

(iii)$\Rightarrow$(i): If $x_1 \geq 0$, $x_2 \geq 0$ solve (2.105), (2.106), then for $x = x_1 - x_2$ and for each $A \in \mathbf{A}$ we have

$$Ax = A(x_1 - x_2) \leq \overline{A}x_1 - \underline{A}x_2 \leq \overline{b}$$

and

$$Ax = A(x_1 - x_2) \geq \underline{A}x_1 - \overline{A}x_2 \geq \underline{b}$$

which shows that $Ax \in \mathbf{b}$ for each $A \in \mathbf{A}$, hence x is a tolerance solution. $\square$

There is a remarkable similarity between the inequality (2.104) and the Oettli–Prager inequality (2.35): both descriptions differ in the sign preceding the matrix Δ only. Yet this seemingly small difference has an astounding impact: although checking the existence of solution of the Oettli–Prager inequality is NP-hard (Theorem 2.12), checking the existence of a tolerance solution can be performed in polynomial time simply by checking solvability of the system (2.105)–(2.107). The description (iii) also shows that the set of tolerance solutions is a convex polyhedron; it allows us to compute the range of components of tolerance solutions by solving the respective linear programming problems [154], etc.

2.17 Control solutions

A vector $x \in \mathbb{R}^n$ is called a *control solution* of $\mathbf{A}x = \mathbf{b}$ if for each $b \in \mathbf{b}$ there exists an $A \in \mathbf{A}$ such that $Ax = b$ holds, in other words, if

$$\mathbf{b} \subseteq \{Ax \mid A \in \mathbf{A}\}.$$

Control solutions were introduced by Shary [178] in 1992. The choice of the word "control" was probably motivated by the fact that each vector $b \in \mathbf{b}$ can be reached by Ax when properly controlling the coefficients of A within $\mathbf{A}$. We have this characterization.

Theorem 2.29. *The following assertions are equivalent.*

(i) x is a control solution of $\mathbf{A}x = \mathbf{b}$.
(ii) x satisfies

$$|A_c x - b_c| \leq \Delta|x| - \delta. \tag{2.109}$$

(iii) x solves

$$A_{ez}x \leq \underline{b}, \tag{2.110}$$
$$-A_{-ez}x \leq -\overline{b} \tag{2.111}$$

for some $z \in Y_n$.

Proof. We prove (i)$\Rightarrow$(ii)$\Rightarrow$(iii)$\Rightarrow$(i).

(i)$\Rightarrow$(ii): If x is a control solution, then by Proposition 2.27 it satisfies $[b_c - \delta, b_c + \delta] \subseteq \{Ax \mid A \in \mathbf{A}\} = [A_c x - \Delta|x|, A_c x + \Delta|x|]$, which implies

$$A_c x - \Delta|x| \leq b_c - \delta \leq b_c + \delta \leq A_c x + \Delta|x|$$

and

$$-(\Delta|x| - \delta) \leq A_c x - b_c \leq \Delta|x| - \delta; \tag{2.112}$$

hence

$$|A_c x - b_c| \leq \Delta|x| - \delta.$$

(ii)$\Rightarrow$(iii): If x satisfies (2.109), then (2.112) holds and with $z = \operatorname{sgn} x$ we can substitute $|x| = T_z x$ into (2.112) which leads to (2.110), (2.111).

(iii)$\Rightarrow$(i): If x solves (2.110), (2.111) for some $z \in Y_n$, then $|\Delta T_z x| \leq \Delta|x|$, hence

$$A_c x - \Delta|x| \leq (A_c - \Delta T_z)x = A_{ez}x \leq \underline{b} \leq \overline{b} \leq A_{-ez}x$$
$$= (A_c + \Delta T_z)x \leq A_c x + \Delta|x|,$$

which implies

$$[\underline{b}, \overline{b}] \subseteq [A_c x - \Delta|x|, A_c x + \Delta|x|] = \{Ax \mid A \in \mathbf{A}\}$$

by Proposition 2.27; hence x is a control solution. $\square$

Again, the inequality (2.109) differs from the Oettli–Prager inequality (2.35) in the sign preceding δ only. But this time the difference does not affect complexity of the problem.

Theorem 2.30. *Checking existence of control solutions is NP-hard.*

Proof. For a square matrix A, consider the system

$$-e \leq Ax \leq e, \tag{2.113}$$

$$e^T|x| \geq 1, \tag{2.114}$$

and the inequality

$$\left| \begin{pmatrix} A \\ 0^T \end{pmatrix} x - \begin{pmatrix} 0 \\ 1 \end{pmatrix} \right| \leq \begin{pmatrix} ee^T \\ e^T \end{pmatrix} |x| - \begin{pmatrix} 0 \\ 0 \end{pmatrix}. \tag{2.115}$$

If x solves (2.113), (2.114), then it also solves (2.115). Conversely, if x solves (2.115), then $x \neq 0$ and $x' = \frac{x}{e^T|x|}$ solves (2.113), (2.114). Hence, the system (2.113), (2.114) has a solution if and only if the inequality (2.115) has a solution. But (2.115) is exactly the inequality (2.109) for the system of interval linear equations

$$[A - ee^T, A + ee^T]x = [0,0], \tag{2.116}$$
$$[-e^T, e^T]x = [1,1], \tag{2.117}$$

which gives that (2.113), (2.114) has a solution if and only if (2.116), (2.117) has a control solution. Now an application of Theorem 2.3 concludes the proof. $\square$

2.18 Algebraic solutions

A vector $x \in \mathbb{R}^n$ is called an *algebraic solution* of $\mathbf{A}x = \mathbf{b}$ if it satisfies

$$\{Ax \mid A \in \mathbf{A}\} = \mathbf{b}. \tag{2.118}$$

Algebraic solutions were first introduced by Ratschek and Sauer in [138]. This type of solution is easy to characterize.

Theorem 2.31. x *is an algebraic solution of* $\mathbf{A}x = \mathbf{b}$ *if and only if it satisfies*

$$A_c x = b_c, \tag{2.119}$$
$$\Delta|x| = \delta. \tag{2.120}$$

Proof. By Proposition 2.27, (2.118) is equivalent to

$$[A_c x - \Delta|x|, A_c x + \Delta|x|] = [b_c - \delta, b_c + \delta], \tag{2.121}$$

which implies (2.119), (2.120). On the other hand, (2.119) and (2.120) imply (2.121) and thus also (2.118). $\square$

It follows from Theorems 2.28 and 2.29, inequalities (2.104) and (2.109), that x is an algebraic solution of $\mathbf{A}x = \mathbf{b}$ if and only if it is both the tolerance and control solution of it. If $m = n$ and A_c is nonsingular, then $\mathbf{A}x = \mathbf{b}$ has an algebraic solution if and only if the data satisfy

$$\Delta|A_c^{-1}b_c| = \delta, \tag{2.122}$$

in which case $x = A_c^{-1}b_c$ is the unique algebraic solution of it.

2.19 The square case

In this section we consider systems of interval linear equations $\mathbf{A}x = \mathbf{b}$ where $\mathbf{A}$ is square $n \times n$ and $\mathbf{b}$ is an n-dimensional interval vector. The square case, which has been a part of the mainstream of interval analysis for the last three decades, would have deserved a special chapter itself, if not a book. Here we confine ourselves to the most important theoretical result (Theorem 2.36), its prerequisites and some of its consequences. Let us repeat that throughout this section $\mathbf{A}$ is square $n \times n$.

As in the noninterval case, nonsingularity plays an important role here. A square interval matrix $\mathbf{A}$ is called *regular* if each $A \in \mathbf{A}$ is nonsingular, and *singular* in the opposite case (i.e., if $\mathbf{A}$ contains a singular matrix). Our previous results imply the following general characterization.

Theorem 2.32. $\mathbf{A}$ *is regular if and only if the system*

$$A_{yc}x^1 - A_{-yc}x^2 = y \tag{2.123}$$

is feasible for each $y \in Y_n$.

Proof. Consider the system of interval linear equations

$$\mathbf{A}x = [-e, e]. \tag{2.124}$$

If $\mathbf{A}$ is regular, then (2.124) is strongly solvable and Theorem 2.14 implies that the system (2.123) is feasible for each $y \in Y_n$ since in this case $b_y = T_y e = y$. Conversely, if (2.123) is feasible for each $y \in Y_n$, then (2.124) is strongly solvable by Theorem 2.14; hence for each $A \in \mathbf{A}$ the system $Ax = e_j$ has a solution for each j, where $e_j \in [-e, e]$ is the jth column of the unit matrix I, which shows that A is invertible and thus nonsingular. $\square$

Theorem 2.33. *Checking regularity of interval matrices is NP-hard.*

Proof. Let A be square. From the proof of Theorem 2.30 we can infer that the system

$$-e \le Ax \le e,$$
$$e^T|x| \ge 1$$

has a solution if and only if the interval matrix

$$[A - ee^T, A + ee^T]$$

is singular. Now Theorem 2.3 provides for the rest.   □

Fortunately, there exists a verifiable sufficient regularity condition that covers most practical cases. $\varrho(A)$ denotes the spectral radius of A.

Proposition 2.34. *If A_c is nonsingular and*

$$\varrho(|A_c^{-1}|\Delta) < 1$$

holds, then $\mathbf{A}$ is regular.

Proof. For each $A \in \mathbf{A}$ we have

$$\varrho(A_c^{-1}(A_c - A)) \leq \varrho(|A_c^{-1}(A_c - A)|) \leq \varrho(|A_c^{-1}|\Delta) < 1.$$

Hence by Theorem 1.31 the matrix

$$I - A_c^{-1}(A_c - A) = A_c^{-1}A$$

is invertible and thus nonsingular. Then A is nonsingular, and $\mathbf{A}$ is regular. □

A square matrix A is called a *P-matrix* if all its principal minors are positive. In 1962 Fiedler and Pták [37] proved this characterization: A is a *P*-matrix if and only if for each $x \neq 0$ there is an i such that $x_i(Ax)_i > 0$ (see Theorem 1.79). With the help of this fact we can prove the next assertion which forms a bridge towards the main result.

Theorem 2.35. *If $\mathbf{A}$ is regular, then $A_1^{-1}A_2$ is a P-matrix for each $A_1, A_2 \in \mathbf{A}$.*

Proof. Assume to the contrary that $A_1^{-1}A_2$ is not a P-matrix for some $A_1, A_2 \in \mathbf{A}$. Then according to the Fiedler–Pták theorem there exists an $x \neq 0$ such that $x_i(A_1^{-1}A_2x)_i \leq 0$ for each i. Take $x' = A_1^{-1}A_2x$; then $x_i x_i' \leq 0$ holds for each i which implies that

$$|x'| + |x| = |x' - x|. \tag{2.125}$$

Now we have

$$|A_c(x' - x)| = |(A_c - A_1)x' + (A_2 - A_c)x| \leq \Delta|x'| + \Delta|x| = \Delta|x' - x|$$

due to (2.125) which also gives that $x' \neq x$ since $x' = x$ would imply $x = 0$ contrary to $x \neq 0$. Hence by the Oettli–Prager theorem there exists an $A \in \mathbf{A}$ with $A(x' - x) = 0$ which means that A is singular, a contradiction. □

When solving an interval linear system $\mathbf{A}x = \mathbf{b}$ with a square interval matrix $\mathbf{A}$, we are usually interested in the set X of weak solutions of it, i.e., in the set

$$X = \{x \mid Ax = b \text{ for some } A \in \mathbf{A}, b \in \mathbf{b}\}. \tag{2.126}$$

The main result of this section asserts that X contains some uniquely determined significant points (Conv X denotes the convex hull of X).

Theorem 2.36. *Let* $\mathbf{A}$ *be regular and let* $\mathbf{b}$ *be an* n-*dimensional interval vector. Then for each* $y \in Y_n$ *the equation*

$$A_c x - T_y \Delta |x| = b_y \tag{2.127}$$

has a unique solution x_y *that belongs to* X *and there holds*

$$\mathrm{Conv}\, X = \mathrm{Conv}\{x_y \mid y \in Y_n\}. \tag{2.128}$$

Proof. Consider the system

$$x^1 = A_{ye}^{-1} A_{-ye} x^2 + A_{ye}^{-1} b_y, \tag{2.129}$$

$$x^1 \geq 0,\ x^2 \geq 0, \tag{2.130}$$

$$(x^1)^T x^2 = 0. \tag{2.131}$$

We can see that (2.129)–(2.131) is a linear complementarity problem [97] whose matrix $A_{ye}^{-1} A_{-ye}$ is a P-matrix due to regularity of $\mathbf{A}$ (Theorem 2.35); hence by the result due to Samelson, Thrall and Wesler [175] (rediscovered independently by Ingleton [55] and Murty [97]), (2.129)–(2.131) has a unique solution x_y^1, x_y^2. Put $x_y = x_y^1 - x_y^2$. Then $A_{ye} x_y^1 - A_{-ye} x_y^2 = b_y$ and Theorem 2.14 gives that for each $A \in \mathbf{A}$ and each $b \in \mathbf{b}$ the unique (because of regularity) solution of $Ax = b$ belongs to $\mathrm{Conv}\{x_y \mid y \in Y_n\}$ which means that $X \subseteq \mathrm{Conv}\{x_y \mid y \in Y_n\}$ and thus also $\mathrm{Conv}\, X \subseteq \mathrm{Conv}\{x_y \mid y \in Y_n\}$. On the other hand, (2.129)–(2.131) imply

$$A_c x_y - b_c = A_c(x_y^1 - x_y^2) - b_c = T_y(\Delta(x_y^1 + x_y^2) + \delta) = T_y(\Delta |x_y| + \delta),$$

so that x_y solves (2.127) and

$$|A_c x_y - b_c| = \Delta |x_y| + \delta \tag{2.132}$$

holds, which in the light of the Oettli–Prager theorem means that $x_y \in X$ for each $y \in Y_n$. Hence $\mathrm{Conv}\{x_y \mid y \in Y_n\} \subseteq \mathrm{Conv}\, X$, which proves the converse inclusion. Finally, if x solves (2.127), then a simple rearrangement shows that $x^1 = x^+$, $x^2 = x^-$ solve (2.129)–(2.131) and in view of the above-stated uniqueness of solution of this linear complementarity problem we have

$$x = x^+ - x^- = x_y^1 - x_y^2 = x_y,$$

so that the solution of (2.127) is unique.   □

Let us emphasize that whereas in Theorem 2.17 x_y denoted an arbitrary of possibly infinitely many solutions of (2.75), in Theorem 2.36 x_y denotes the unique solution of (2.127). If $\mathbf{A}$ is regular, then for each $y \in Y_n$ the point x_y can be computed by the following finite algorithm, called the sign-accord algorithm because it works towards achieving a "sign accord" of vectors z and x (i.e., $z_j x_j \geq 0$ for each j).

$$
\begin{array}{|l|}
\hline
z := \mathrm{sgn}\,(A_c^{-1} b_y); \\
x := A_{yz}^{-1} b_y; \\
C := A_{yz}^{-1} T_y \Delta; \\
\text{\textbf{while} } z_j x_j < 0 \text{ for some } j \\
\quad k := \min\{j \mid z_j x_j < 0\}; \\
\quad z_k := -z_k; \\
\quad \alpha := 2z_k/(1 - 2z_k C_{kk}); \\
\quad x := x + \alpha x_k C_{\cdot k}; \\
\quad C := C + \alpha C_{\cdot k} C_{k \cdot}; \\
\text{\textbf{end}} \\
x_y := x. \\
\hline
\end{array}
$$

($C_{\cdot k}$ and $C_{k \cdot}$ denote the kth column and the kth row of C, respectively.) We refrain from including a proof here, which would lead us beyond the scope of this chapter. We refer an interested reader to [156, p. 48].

The narrowest interval vector $[\underline{x}, \overline{x}]$ containing the set X is called the *interval hull* of X. From (2.128) we immediately have that

$$\underline{x}_i = \min_{y \in Y_n} (x_y)_i, \tag{2.133}$$

$$\overline{x}_i = \max_{y \in Y_n} (x_y)_i \tag{2.134}$$

($i = 1, \ldots, n$), which, when combined with the sign-accord algorithm, yields a finite procedure for computing the interval hull.

Example 2.37. Consider the example by Hansen [47]: $\mathbf{A} = [\underline{A}, \overline{A}]$, $\mathbf{b} = [\underline{b}, \overline{b}]$, where

$$\underline{A} = \begin{pmatrix} 2 & 0 \\ 1 & 2 \end{pmatrix}, \quad \overline{A} = \begin{pmatrix} 3 & 1 \\ 2 & 3 \end{pmatrix}, \quad \underline{b} = \begin{pmatrix} 0 \\ 60 \end{pmatrix}, \quad \overline{b} = \begin{pmatrix} 120 \\ 240 \end{pmatrix}.$$

Since for each $z \in Y_2$ the intersection of the set of weak solutions X with the orthant $\{x \in \mathbb{R}^2 \mid T_z x \geq 0\}$ is described by the system of linear inequalities (2.39), (2.40), considering separately all four orthants we arrive at this picture of the set X:

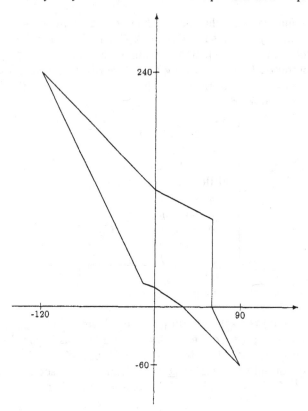

It can be seen that X is nonconvex and the four points x_y, $y \in Y_2$ are clearly visible since in view of (2.128) they must be exactly the four vertices of the convex hull of X. Using the sign-accord algorithm, we obtain

$$x_{(-1,-1)} = (-12, 24)^T,$$
$$x_{(-1,1)} = (-120, 240)^T,$$
$$x_{(1,-1)} = (90, -60)^T,$$
$$x_{(1,1)} = (60, 90)^T,$$

and from (2.133), (2.134) we have that the interval hull of X is $[\underline{x}, \overline{x}]$, where

$$\underline{x} = (-120, -60)^T,$$
$$\overline{x} = (90, 240)^T.$$

Unfortunately, the general problem is again NP-hard:

Theorem 2.38. *Computing the interval hull of the set X is NP-hard even for systems with interval matrices satisfying*

$$\varrho(|A_c^{-1}|\Delta) = 0.$$

Proof. Given a rational $n \times n$ matrix A, construct the $(n+1) \times (n+1)$ interval matrix $\mathbf{A} = [A_c - \Delta, A_c + \Delta]$ with

$$A_c = \begin{pmatrix} 1 & 0^T \\ 0 & A \end{pmatrix}, \quad \Delta = \begin{pmatrix} 0 & e^T \\ 0 & 0 \end{pmatrix},$$

and the $(n+1)$-dimensional interval vector $\mathbf{b} = [b_c - \delta, b_c + \delta]$ with

$$b_c = \begin{pmatrix} 0 \\ 0 \end{pmatrix}, \quad \delta = \begin{pmatrix} 0 \\ e \end{pmatrix}$$

$(e \in \mathbb{R}^n)$. We have

$$|A_c^{-1}|\Delta = \begin{pmatrix} 0 & e^T \\ 0 & 0 \end{pmatrix};$$

hence

$$\varrho(|A_c^{-1}|\Delta) = 0.$$

Then each system $Ax = b$ with $A \in \mathbf{A}$, $b \in \mathbf{b}$ has the form

$$x_1 + c^T x' = 0,$$
$$Ax' = d$$

for some $c \in [-e, e]$ and $d \in [-e, e]$, where $x' = (x_2, \ldots, x_{n+1})^T$. If $[\underline{x}, \overline{x}]$ is the interval hull of (2.126), then for $\overline{x}_1$ we have

$$\overline{x}_1 = \max\{c^T x' \mid c \in [-e, e], -e \le Ax' \le e\} = \max\{e^T |x'| \mid -e \le Ax' \le e\};$$

hence

$$\overline{x}_1 \ge 1$$

holds if and only if the system

$$-e \le Ax' \le e,$$

$$e^T |x'| \ge 1$$

has a solution. Since the latter problem is NP-complete (Theorem 2.3), $\overline{x}_1$ is NP-hard to compute and the same holds for $[\underline{x}, \overline{x}]$. $\square$

This result shows that we must set the goal differently: instead of trying to compute the exact interval hull $[\underline{x}, \overline{x}]$, we should be satisfied with computing a possibly narrow *enclosure* of X, i.e., an interval vector $[\underline{x}, \overline{\overline{x}}]$ satisfying

$$X \subseteq [\underline{x}, \overline{\overline{x}}].$$

There is a vast literature dedicated to this theme, comprising a number of ingenious enclosure methods, see, e.g., the monographs by Alefeld and Herzberger [2] or Neumaier [105]. We conclude this section with description of a nontrivial result that gives explicit formulae for computing an enclosure.

Theorem 2.39 (Hansen–Bliek–Rohn). *Let A_c be nonsingular and let*

$$\varrho(|A_c^{-1}|\varDelta) < 1 \tag{2.135}$$

hold. Then we have

$$X \subseteq [\min\{\underset{\sim}{x}, T_\nu \underset{\sim}{x}\}, \max\{\tilde{x}, T_\nu \tilde{x}\}], \tag{2.136}$$

where

$$
\begin{aligned}
M &= (I - |A_c^{-1}|\varDelta)^{-1}, \\
\mu &= (M_{11}, \ldots, M_{nn})^T, \\
T_\nu &= (2T_\mu - I)^{-1}, \\
x_c &= A_c^{-1}b_c, \\
x^* &= M(|x_c| + |A_c^{-1}|\delta), \\
\underset{\sim}{x} &= -x^* + T_\mu(x_c + |x_c|), \\
\tilde{x} &= \quad x^* + T_\mu(x_c - |x_c|).
\end{aligned}
$$

Proof. First we note that because of (2.135) we have

$$M = \sum_{j=0}^{\infty}(|A_c^{-1}|\varDelta)^j \geq I \geq 0;$$

thus also $2T_\mu - I \geq I$, so that the diagonal matrix $T_\nu = (2T_\mu - I)^{-1}$ exists and $\nu_i = 1/(2M_{ii} - 1)$ for each i.

To prove (2.136), take an $x \in X$; then by the Oettli–Prager theorem it satisfies

$$|A_c x - b_c| \leq \varDelta|x| + \delta;$$

hence

$$|x| - |x_c| \leq |x - x_c| = |A_c^{-1}(A_c x - b_c)| \leq |A_c^{-1}||A_c x - b_c| \leq |A_c^{-1}|(\varDelta|x| + \delta). \tag{2.137}$$

Now, let us fix an $i \in \{1, \ldots, n\}$. Then from (2.137) we have

$$x_i \leq (x_c)_i + (|A_c^{-1}|(\varDelta|x| + \delta))_i \tag{2.138}$$

and

$$|x_j| \leq |x_c|_j + (|A_c^{-1}|(\varDelta|x| + \delta))_j \tag{2.139}$$

for each $j \neq i$. Since $x_i = |x_i| + (x_i - |x_i|)$ and the same holds for $(x_c)_i$, we can put (2.138) and (2.139) together as

$$|x| + (x_i - |x_i|)e_i \leq |x_c| + ((x_c)_i - |x_c|_i)e_i + |A_c^{-1}|(\varDelta|x| + \delta),$$

which implies

$$(I - |A_c^{-1}|\Delta)|x| + (x_i - |x_i|)e_i \leq |x_c| + |A_c^{-1}|\delta + ((x_c)_i - |x_c|_i)e_i.$$

Premultiplying this inequality by the nonnegative vector $e_i^T M$, we finally obtain an inequality containing variable x_i only:

$$|x_i| + (x_i - |x_i|)M_{ii} \leq x_i^* + ((x_c)_i - |x_c|_i)M_{ii} = \tilde{x}_i.$$

If $x_i \geq 0$, then this inequality becomes

$$x_i \leq \tilde{x}_i,$$

and if $x_i < 0$, then it turns into

$$x_i \leq \tilde{x}_i/(2M_{ii} - 1) = \nu_i \tilde{x}_i,$$

in both cases

$$x_i \leq \max\{\tilde{x}_i, \nu_i \tilde{x}_i\}.$$

Since i was arbitrary, we conclude that

$$x \leq \max\{\tilde{x}, T_\nu \tilde{x}\},$$

which is the upper bound in (2.136). To prove the lower bound, notice that if $Ax = b$ for some $A \in \mathbf{A}$ and $b \in \mathbf{b}$, then $A(-x) = -b$, hence $-x$ belongs to the solution set of the system $\mathbf{A}x = [-b_c - \delta, -b_c + \delta]$, and we can apply the previous result to this system by setting $b_c := -b_c$. In this way we obtain

$$-x \leq \max\{x^* + T_\mu(-x_c - |x_c|), T_\nu(x^* + T_\mu(-x_c - |x_c|))\};$$

hence

$$x \geq \min\{-x^* + T_\mu(x_c + |x_c|), T_\nu(-x^* + T_\mu(x_c + |x_c|))\} = \min\{\underline{x}, T_\nu \underline{x}\},$$

which is the lower bound in (2.136). The theorem is proved. □

This theorem gives an enclosure (2.136) which is fairly good in practical cases, but generally not optimal (cf. Theorem 2.38). However, it is optimal (i.e., it yields the interval hull of X) in the case of $A_c = I$ (Hansen [48], Bliek [21], Rohn [160]).

The result can be easily applied to bound the inverse of an interval matrix. In the next theorem, the minimum or maximum of two matrices is understood componentwise.

Theorem 2.40. *Let (2.135) hold. Then for each $A \in \mathbf{A}$ we have*

$$\min\{\underline{B}, T_\nu \underline{B}\} \leq A^{-1} \leq \max\{\widetilde{B}, T_\nu \widetilde{B}\},$$

where M, μ, and T_ν are as in Theorem 2.39 and

$$\underline{B} = -M|A_c^{-1}| + T_\mu(A_c^{-1} + |A_c^{-1}|),$$

$$\widetilde{B} = M|A_c^{-1}| + T_\mu(A_c^{-1} - |A_c^{-1}|).$$

Proof. Since $(A^{-1})_{.j}$ is the solution of the system $Ax = e_j$, we obtain the result simply by applying Theorem 2.39 to interval linear systems $\mathbf{A}x = [e_j, e_j]$ for $j = 1, \ldots, n$. □

2.20 Summary II: Solution types

We have introduced altogether eight types of solutions. We summarize the results in the following table which clearly illustrates the tiny differences in their descriptions.

Solution	Description	Reference
weak solution of $\mathbf{A}x = \mathbf{b}$	$\|A_c x - b_c\| \le \Delta\|x\| + \delta$	(2.35)
strong solution of $\mathbf{A}x = \mathbf{b}$	$A_c x - b_c = \Delta\|x\| = \delta = 0$	(2.68), (2.69)
weak solution of $\mathbf{A}x \le \mathbf{b}$	$A_c x - b_c \le \Delta\|x\| + \delta$	(2.82)
strong solution of $\mathbf{A}x \le \mathbf{b}$	$A_c x - b_c \le -\Delta\|x\| - \delta$	(2.94)
tolerance solution	$\|A_c x - b_c\| \le -\Delta\|x\| + \delta$	(2.104)
control solution	$\|A_c x - b_c\| \le \Delta\|x\| - \delta$	(2.109)
algebraic solution	$A_c x - b_c = \Delta\|x\| - \delta = 0$	(2.119), (2.120)
x_y	$\|A_c x - b_c\| = \Delta\|x\| + \delta$	(2.132)

2.21 Notes and references

In this section we give some additional notes and references to the material contained in this chapter.

Section 2.1. We use standard linear algebraic notations except for Y_m, T_y and sgn x (introduced in [156]).

Section 2.2. The algorithm is a variant of the binary reflected Gray code (Gray [46]), see, e.g., [194].

Section 2.3. The first NP-hardness result for problems with interval-valued data was published by Poljak and Rohn as a report [116] in 1988 and as a journal paper [117] in 1993. They showed that for an $n \times n$ matrix A the value

$$\max_{z,y \in Y_n} z^T A y \tag{2.140}$$

is NP-hard to compute, and they used the result to prove that checking regularity of interval matrices is NP-hard (Theorem 2.33 here). Only in 1995 was it realized [162] that the value of (2.140) is equal to $\|A\|_{\infty,1}$ (see (2.4)) which led to the formulation of Theorem 2.2 ([162], in journal form [165]). Theorem 2.3, which is more useful in the context of interval linear systems, was also proved in [162]. Notice that all the NP-hardness results of this chapter were proved with the help of this theorem.

Section 2.4. The word "feasibility", which is a one-word substitute for non-negative solvability, was inspired by linear programming terminology. Theorem 2.4, also known as Farkas' lemma, was proved by Farkas [34] in 1902. It is an important theoretical result (as evidenced throughout this chapter), but it does not give a constructive way of checking feasibility which must be done by another means (usually by a linear programming technique).

Section 2.5. Matrices A_{yz} and vectors b_y were introduced in [156]. The importance of the finite set of matrices A_{yz} becomes more apparent with problems involving square interval matrices only (as regularity, positive definiteness etc.). For example, an interval matrix $\mathbf{A}$ is regular (see Section 2.19) if and only if $\det(A_{yz})$ is of the same sign for each $z, y \in Y_n$ (Baumann [10]); for further results of this type see the monograph by Kreinovich, Lakeyev, Rohn, and Kahl [80, Chapters 21 and 22]. As we have seen, in the context of rectangular interval systems typically only matrices of the form A_{ye} or A_{ez} arise.

Section 2.6. The definition of an interval linear system $\mathbf{A}x = \mathbf{b}$ as a family of systems $Ax = b$, $A \in \mathbf{A}$, $b \in \mathbf{b}$ makes it possible to define various types of solutions. The notion of strong feasibility of interval linear equations was introduced in [149], and weak solvability as a counterpart of strong solvability was first studied by Rohn and Kreslová in [168]. Formulation and study of the complete set of the eight decision problems is new and forms the bulk of this chapter.

Section 2.7. The Oettli–Prager theorem is formulated here in the form (2.35) which has become standard, although not explicitly present in the original paper [112] where the authors preferred an entrywise formulation. The theorem is now considered a basic tool for both backward error analysis (Golub and van Loan [44], Higham [51]) and interval analysis (Neumaier [105]) of systems of linear equations. Another form of Proposition 2.10 (perhaps more attractive, but less useful) is described in [153, Theorem 1.2]. NP-hardness of checking weak solvability of equations was proved by Lakeyev and Noskov [84] (preliminary announcement without proof in [83]) by another means. The proof given here employs polynomial reduction of our standard problem of Theorem 2.3 to the current problem, an approach adhered to throughout the chapter.

Section 2.8. Theorem 2.13 is a simple consequence of the Oettli–Prager theorem. It was discovered independently in [145].

Section 2.9. The proof of Theorem 2.14 is not straightforward and neither is its history. The "if" part was formulated and proved in technical reports [152], [151] in 1984, but the author refrained from further journal publication because he considered the sufficient condition too strong. In 1996 he discovered by chance that it was also necessary (paradoxically, it was the easier part of the proof), which gave rise to Theorem 2.14 published in [166]. The second part of the proof of the "if" part (starting from (2.56)) relies in fact on a new existence theorem for systems of linear equations which was published in [157] (existence proof, as given here) and in [159] (constructive proof). NP-hardness of checking strong solvability (Theorem 2.15) is an easy consequence of the same complexity result for the problem of checking regularity of interval matrices (Theorem 2.33), but because of the layout of this chapter it had to be proved independently.

Section 2.10. Characterization of strong feasibility of equations (Theorem 2.17) was published in [149] as part of a study of the interval linear programming problem. Many unsuccessful attempts by the author through the following years to find a characterization of strong feasibility that would not be inherently exponential finally led to the NP-hardness conjecture and to the proof of it in [164] (part 2 of the proof).

Section 2.11. Gerlach [43] initiated the study of systems of interval linear inequalities by proving Theorem 2.19 as a follow-up of the Oettli–Prager theorem. NP-hardness of checking weak solvability of inequalities was proved in technical report [162] and has not been published in journal form.

Section 2.12. The result of Theorem 2.22 is obvious and is included here for completeness.

Section 2.13. Both Theorems 2.23 and 2.24 are due to Rohn and Kreslová [168]. The contrast between the complexity results for strong solvability of interval linear equations (Theorem 2.15) and inequalities (Theorem 2.23) is striking and reveals that classical solvability-preserving reductions between linear equations and linear inequalities are no longer in force when inexact data are present. In fact, a system of linear equations $Ax = b$ can be equivalently written as a system if linear inequalities $Ax \leq b$, $-Ax \leq -b$ and solved as such. But in the case of interval data, the sets of weak solutions of $\mathbf{A}x = \mathbf{b}$ and of $\mathbf{A}x \leq \mathbf{b}$, $-\mathbf{A}x \leq -\mathbf{b}$ are generally not identical since the latter family contains systems of inequalities of type $Ax \leq b$, $-\tilde{A}x \leq -\tilde{b}$ ($A, \tilde{A} \in \mathbf{A}$, $b, \tilde{b} \in \mathbf{b}$) that may possess solutions which do not satisfy $Ax = b$ for any $A \in \mathbf{A}$, $b \in \mathbf{b}$. Existence of strong solutions in the case of strong solvability (Theorem 2.24) is a nontrivial fact that can be expected to find some applications, although none of them have been known to date.

Section 2.14. Theorem 2.26 is again obvious.

Section 2.16. Introduction of the notion of tolerance solutions was motivated by considerations concerning crane construction (Nuding and Wilhelm [110]) and input–output planning with inexact data of the socialist economy of the former Czechoslovakia [146]. Descriptions (ii), (iii) of tolerance solutions in Theorem 2.28 were proved in [154]. Tolerance solutions have been studied since by Neumaier [104], Deif [32], Kelling and Oelschlägel [70], Kelling [68], [69], Shaydurov and Shary [185], Shary [176], [179], [180], [181], and Lakeyev and Noskov [84].

Section 2.17. Control solutions were introduced by Shary [178] and further studied by him in [181], [183]. The description (2.109) in Theorem 2.29 is due to Lakeyev and Noskov [84] who in the same paper also proved NP-hardness of checking the existence of control solutions, as well as of algebraic solutions. For other possible types of solutions see the survey paper by Shary [184].

Section 2.18. Algebraic solutions were introduced by Ratschek and Sauer [138], although for the case $m = 1$ only. The condition (2.122) was proved in [153]. The topic makes more sense when the problem is formulated as $\mathbf{A} \cdot \mathbf{x} = \mathbf{b}$, where $\mathbf{x}$ is an interval vector and multiplication is performed in interval arithmetic. A solution of this problem in full generality is not known so far; for a partial solution see [158].

Section 2.19. NP-hardness of checking regularity of square matrices (Theorem 2.33) was proved by Poljak and Rohn [116], [117] whose work was motivated by the existence at that time of more than ten necessary and sufficient regularity conditions all of which exhibited exponential complexity (Theorem 5.1 in [156]; one of them is our Theorem 2.32). The sufficient regularity condition of Proposition 2.34 is usually attributed to Beeck [11], although allegedly (Neumaier [103]) it was derived earlier by Ris in his unpublished Ph.D. thesis [143]. The "convex-hull" Theorem 2.36, as well as finiteness of the sign accord algorithm, were proved in [156]. The NP-hardness result of Theorem 2.38 on complexity of computing the interval hull of the set X of weak solutions is due to Rohn and Kreinovich [167]. In 1992 Hansen [48] and Bliek [21] showed almost simultaneously that in the case $A_c = I$ the interval hull can be described by closed-form formulae, but their result lacked a rigorous proof which was supplied in [160]. The idea can be applied to a preconditioned system, as was done in the proof of Theorem 2.39, but in this way only an enclosure, not the interval hull, is obtained (Theorem 2.38 explains why it is so). Computation of the enclosure requires evaluation of two inverses, A_c^{-1} and $(I - |A_c^{-1}|\Delta)^{-1}$; the main result of [144] shows that we can also do with approximate inverses $R \approx A_c^{-1}$ and $M \approx (I - |A_c^{-1}|\Delta)^{-1}$ provided they satisfy certain additional inequality. The topic was later studied by Ning and Kearfott [108] and Neumaier [106]. *We refrain here from listing papers dedicated to computing enclosures since they are simply too many.* As for the latest developments,[1] we mention the method of Jansson [64], characterization of feasibility of preconditioned interval Gaussian algorithm by Mayer and Rohn [87], the techniques by Shary [177], [182], and a series of papers by Alefeld, Kreinovich, and Mayer [6], [3], [4], [5] which handle the complicated problem of solving interval systems with dependent data. An earlier version of this problem (with prescribed bounds on column sums of $A \in \mathbf{A}$) was studied in [148].

Works related to the material of this chapter include (but are not limited to) Albrecht [1], Coxson [28], Garloff [42], Heindl [49], Herzberger and Bethke [50], Jahn [60], Moore [89], [90], Nedoma [99], [100], [101], [102], Nickel [107], Nuding [109], Oettli [111], Rex [140], Rex and Rohn [141], [142], Rump [171], [172], [173] and Shokin [186], [187].

---

[1]Written in Spring 2002.

3

Interval linear programming

J. Rohn

3.1 Linear programming: Duality

We now switch to optimization problems. Given $A \in \mathbb{R}^{m \times n}$, $b \in \mathbb{R}^m$ and $c \in \mathbb{R}^n$, the problem

$$\text{minimize } c^T x \qquad (3.1)$$

subject to (s.t.)

$$Ax = b, \, x \geq 0 \qquad (3.2)$$

is called a linear programming problem, or simply a linear program. We write the problem (3.1), (3.2) briefly as

$$\text{Min}\{c^T x \mid Ax = b, x \geq 0\} \qquad (3.3)$$

(notice the use of the upper case in "Min" to denote a problem in contrast to "min" which denotes minimum when applicable). A vector x satisfying (3.2) is called a *feasible solution* of (3.3). A problem (3.3) having a feasible solution is said to be *feasible*, and *infeasible* in the opposite case. Hence, the problem (3.3) is feasible if and only if the system $Ax = b$ is feasible in the terminology of Section 2.4.

For a given linear program (3.3) we introduce the value

$$f(A, b, c) = \inf\{c^T x \mid Ax = b, x \geq 0\} \qquad (3.4)$$

and we call it the *optimal value* of (3.3).[1] The optimal value can be computed by any linear programming technique, such as, e.g., the simplex method by Dantzig [31], or the polynomial-time algorithms by Khachiyan [71], Karmarkar [67] and others (see Padberg [115]). Exactly one of the following three cases may occur.

---

[1]In linear programming only the finite value of $f(A, b, c)$ is accepted as the optimal value; we use this formulation for the sake of utmost generality of later results.

(a) If $f(A, b, c)$ is finite, then, as proved in part (a) of the proof of Theorem 3.1 below, the infimum in (3.4) is attained as minimum, so that there exists a feasible solution x^* of (3.3) satisfying $f(A, b, c) = c^T x^*$. Such an x^* is called an *optimal solution* of (3.3). In this case we say that the problem (3.3) has an optimal solution.

(b) If $f(A, b, c) = -\infty$, then the set of feasible solutions of (3.3) contains a half-line along which the value of $c^T x$ tends to $-\infty$ (see part (b) of the proof of Theorem 3.1); in this case we call the problem (3.3) *unbounded*.

(c) If $f(A, b, c) = \infty$, then the set of feasible solutions of (3.3) is empty; hence the problem (3.3) is infeasible.

Given a problem (3.3) (called "*primal*" in this context), we can formulate its *dual problem* as

$$\text{maximize } b^T p \tag{3.5}$$

s.t.

$$A^T p \le c, \tag{3.6}$$

or briefly

$$\text{Max}\{b^T p \mid A^T p \le c\} \tag{3.7}$$

(notice that the nonnegativity constraint is missing in (3.6)). The dual problem is called *solvable* if the system $A^T p \le c$ is solvable,[2] and *unsolvable* in the opposite case. In analogy with the primal problem, we introduce for the dual problem the value

$$g(A, b, c) = \sup\{b^T p \mid A^T p \le c\}.$$

A solution p^* of $A^T p \le c$ is called an optimal solution of (3.7) if $g(A, b, c) = b^T p^*$; if $g(A, b, c) = -\infty$, then the problem (3.7) is unsolvable, and if $g(A, b, c) = \infty$, then it is called *unbounded*. The primal and the dual problem are connected by the following important result whose proof is included here for the sake of completeness.

Theorem 3.1 (Duality theorem). *If $f(A, b, c) < \infty$ or $g(A, b, c) > -\infty$, then*

$$f(A, b, c) = g(A, b, c). \tag{3.8}$$

Comment. The formal equality (3.8) covers three qualitative issues: (i) if one of the problems (3.3), (3.7) has an optimal solution, then so does the second one and the optimal values of both problems are equal; (ii) if the primal problem (3.3) is unbounded, then the dual problem (3.7) is unsolvable; (iii) if the dual problem (3.7) is unbounded, then the primal problem (3.3) is infeasible. If the assumptions of the theorem are not met, then (3.3) is infeasible and (3.7) is unsolvable, in which case nothing more can be said.

---

[2] At this point we must depart from traditional linear programming terminology where (3.7) is called feasible if $A^T p \le c$ has a solution; but for us feasibility means nonnegative solvability, so that we cannot use this term here and we must stick to terminology introduced in Section 2.4.

Proof. Three possibilities may occur under our assumptions: (a) $f(A, b, c) < \infty$ and $g(A, b, c) > -\infty$, (b) $f(A, b, c) < \infty$ and $g(A, b, c) = -\infty$, and (c) $f(A, b, c) = \infty$ and $g(A, b, c) > -\infty$.

(a) Let $f(A, b, c) < \infty$ and $g(A, b, c) > -\infty$. We first prove that the system

$$Ax = b, \; x \geq 0, \tag{3.9}$$

$$A^T p \leq c, \tag{3.10}$$

$$c^T x \leq b^T p \tag{3.11}$$

has a solution. Introducing artificial variables, we can write it in the form

$$\begin{pmatrix} A & 0 & 0 & 0 & 0 \\ 0 & A^T & -A^T & I & 0 \\ c^T & -b^T & b^T & 0 & 1 \end{pmatrix} \begin{pmatrix} x \\ p_1 \\ p_2 \\ p_3 \\ \xi \end{pmatrix} = \begin{pmatrix} b \\ c \\ 0 \end{pmatrix}, \tag{3.12}$$

where all the variables are nonnegative. Now we can apply the Farkas theorem which says that (3.12) has a nonnegative solution if and only if for each t_1, t_2 and τ,

$$\begin{pmatrix} A^T & 0 & c \\ 0 & A & -b \\ 0 & -A & b \\ 0 & I & 0 \\ 0 & 0 & 1 \end{pmatrix} \begin{pmatrix} t_1 \\ t_2 \\ \tau \end{pmatrix} \geq 0 \quad \text{implies} \quad \begin{pmatrix} b \\ c \\ 0 \end{pmatrix}^T \begin{pmatrix} t_1 \\ t_2 \\ \tau \end{pmatrix} \geq 0, \tag{3.13}$$

which means that

$$A^T t_1 + \tau c \geq 0, \tag{3.14}$$

$$A t_2 = \tau b, \tag{3.15}$$

$$t_2 \geq 0, \; \tau \geq 0 \tag{3.16}$$

should imply

$$b^T t_1 + c^T t_2 \geq 0. \tag{3.17}$$

To prove the last statement, in view of the nonnegativity of τ we can consider two cases. If $\tau > 0$, then we have $b = \frac{1}{\tau} A t_2$; hence

$$b^T t_1 + c^T t_2 = \tfrac{1}{\tau} t_2^T A^T t_1 + c^T t_2 = \tfrac{1}{\tau} t_2^T (A^T t_1 + \tau c) \geq 0$$

because of (3.14), (3.16), which is (3.17). If $\tau = 0$, then (3.14)–(3.16) turn into

$$A^T t_1 \geq 0, \tag{3.18}$$

$$A t_2 = 0, \tag{3.19}$$

$$t_2 \geq 0. \tag{3.20}$$

Since $f(A, b, c) < \infty$, the system $Ax = b$ is feasible and (3.18) by the Farkas theorem implies $b^T t_1 \geq 0$; since $g(A, b, c) > -\infty$, the system $A^T p \leq c$ is solvable and (3.19), (3.20) by Theorem 2.6 imply $c^T t_2 \geq 0$. Hence (3.17) again holds. In this way we have proved the implication (3.13), which in turn guarantees existence of a solution x^*, p^* of the system (3.9)–(3.11). From (3.9), (3.10) we obtain

$$c^T x^* = x^{*T} c \geq x^{*T} A^T p^* = (Ax^*)^T p^* = b^T p^*,$$

which together with (3.11) gives $c^T x^* = b^T p^*$. Summing up, we have proved that there exist x^*, p^* satisfying

$$Ax^* = b, \; x^* \geq 0,$$

$$A^T p^* \leq c,$$

$$c^T x^* = b^T p^*.$$

Now, for each feasible solution x of the primal problem we have

$$c^T x = x^T c \geq x^T A^T p^* = (Ax)^T p^* = b^T p^* = c^T x^*,$$

which means that

$$c^T x^* = \min\{c^T x \mid Ax = b, x \geq 0\} = f(A, b, c),$$

and similarly for each solution p of the system of constraints $A^T p \leq c$ of the dual problem we have

$$b^T p = p^T b = p^T Ax^* = (A^T p)^T x^* \leq c^T x^* = b^T p^*,$$

which gives that

$$b^T p^* = \max\{b^T p \mid A^T p \leq c\} = g(A, b, c),$$

and finally

$$f(A, b, c) = c^T x^* = b^T p^* = g(A, b, c),$$

which is (3.8).

(b) Let $f(A, b, c) < \infty$ and $g(A, b, c) = -\infty$. Then the primal problem has a feasible solution, say x_1, and the dual problem is unsolvable, so that the system $A^T p \leq c$ has no solution; hence according to Theorem 2.6 there exists an x_0 satisfying $Ax_0 = 0$, $x_0 \geq 0$ and $c^T x_0 < 0$. Then for each $\alpha \in \mathbb{R}$, $\alpha \geq 0$ we have $A(x_1 + \alpha x_0) = Ax_1 = b$ and $x_1 + \alpha x_0 \geq 0$; hence $x_1 + \alpha x_0$ is a feasible solution of the primal problem for each $\alpha \geq 0$ and

$$\lim_{\alpha \to \infty} c^T (x_1 + \alpha x_0) = \lim_{\alpha \to \infty} (c^T x_1 + \alpha c^T x_0) = -\infty$$

because of $c^T x_0 < 0$; hence

$$f(A, b, c) = \inf\{c^T x \mid Ax = b, x \geq 0\} = -\infty = g(A, b, c),$$

which is (3.8).

(c) Let $f(A, b, c) = \infty$ and $g(A, b, c) > -\infty$. Then the primal problem is infeasible and the system of constraints $A^T p \leq c$ of the dual problem has a solution, say p_1. Since the system $Ax = b$ is not feasible, according to the Farkas theorem there exists a p_0 satisfying $A^T p_0 \geq 0$ and $b^T p_0 < 0$. Then for each $\alpha \in \mathbb{R}$, $\alpha \geq 0$ we have $A^T(p_1 - \alpha p_0) \leq A^T p_1 \leq c$ and

$$\lim_{\alpha \to \infty} b^T(p_1 - \alpha p_0) = \lim_{\alpha \to \infty} (b^T p_1 - \alpha b^T p_0) = \infty$$

because of $b^T p_0 < 0$; hence

$$g(A, b, c) = \sup\{b^T p \mid A^T p \leq c\} = \infty = f(A, b, c),$$

which is (3.8).

We have proved that in all three cases (a), (b), (c) the equality (3.8) holds. This concludes the proof. $\qquad\square$

3.2 Interval linear programming problem

Let $\mathbf{A} = [\underline{A}, \overline{A}] = [A_c - \Delta, A_c + \Delta]$ be an $m \times n$ interval matrix and let $\mathbf{b} = [\underline{b}, \overline{b}] = [b_c - \delta, b_c + \delta]$ and $\mathbf{c} = [\underline{c}, \overline{c}] = [c_c - \gamma, c_c + \gamma]$ be an m-dimensional and n-dimensional interval vector, respectively. The *family* of linear programming problems

$$\text{Min}\{c^T x \mid Ax = b, x \geq 0\} \tag{3.21}$$

with data satisfying

$$A \in \mathbf{A},\ b \in \mathbf{b},\ c \in \mathbf{c} \tag{3.22}$$

is called an *interval linear programming problem*. Since for each linear programming problem (3.21) we have a uniquely determined optimal value $f(A, b, c)$, it is natural to consider its range over the data (3.22) by introducing the values

$$\underline{f}(\mathbf{A}, \mathbf{b}, \mathbf{c}) = \inf\{f(A, b, c) \mid A \in \mathbf{A}, b \in \mathbf{b}, c \in \mathbf{c}\},$$

$$\overline{f}(\mathbf{A}, \mathbf{b}, \mathbf{c}) = \sup\{f(A, b, c) \mid A \in \mathbf{A}, b \in \mathbf{b}, c \in \mathbf{c}\}.$$

The interval $[\underline{f}(\mathbf{A}, \mathbf{b}, \mathbf{c}), \overline{f}(\mathbf{A}, \mathbf{b}, \mathbf{c})]$, whose bounds may be infinite, is called the *range of the optimal value* of the interval linear programming problem (3.21), (3.22). In the next section we derive formulae for computing the range that are the cornerstone of our approach.

3.3 Range of the optimal value

The following theorem gives explicit formulae for computing the bounds of the range. Notice that the result holds without any additional assumptions.

Theorem 3.2. *We have*

$$\underline{f}(\mathbf{A}, \mathbf{b}, \mathbf{c}) = \inf\{\underline{c}^T x \mid \underline{A}x \leq \overline{b}, \overline{A}x \geq \underline{b}, x \geq 0\}, \tag{3.23}$$

$$\overline{f}(\mathbf{A}, \mathbf{b}, \mathbf{c}) = \sup_{y \in Y_m} f(A_{ye}, b_y, \overline{c}). \tag{3.24}$$

Comment. Hence, solving only one linear programming problem is needed to evaluate $\underline{f}(\mathbf{A}, \mathbf{b}, \mathbf{c})$, whereas up to 2^m of them are to be solved to compute $\overline{f}(\mathbf{A}, \mathbf{b}, \mathbf{c})$ according to (3.24). Although the set Y_m is finite, we use "sup" here because some of the values may be infinite.

Proof. For given $\mathbf{A}$, $\mathbf{b}$, $\mathbf{c}$ denote $\underline{f} := \underline{f}(\mathbf{A}, \mathbf{b}, \mathbf{c})$, $\overline{f} := \overline{f}(\mathbf{A}, \mathbf{b}, \mathbf{c})$.
   (a) To prove (3.23), put

$$\varphi = \inf\{\underline{c}^T x \mid \underline{A}x \leq \overline{b}, \overline{A}x \geq \underline{b}, x \geq 0\}.$$

   (a.1) First we prove $\underline{f} \leq \varphi$. This is obvious if $\varphi = \infty$. If $\varphi < \infty$, then the linear system

$$\underline{A}x \leq \overline{b}, \ \overline{A}x \geq \underline{b} \tag{3.25}$$

is feasible. Let x be a nonnegative solution of it. Then in view of Theorem 2.13, x is a nonnegative weak solution of $Ax = b$; hence there exist $A \in \mathbf{A}$, $b \in \mathbf{b}$ such that $Ax = b$ holds. Then $\underline{f} \leq f(A, b, \underline{c}) \leq \underline{c}^T x$, and since x is an arbitrary nonnegative solution of (3.25), we obtain $\underline{f} \leq \varphi$.
   (a.2) Second we prove $\varphi \leq \underline{f}$ by showing that

$$\varphi \leq f(A, b, c) \tag{3.26}$$

holds for each $A \in \mathbf{A}$, $b \in \mathbf{b}$, $c \in \mathbf{c}$. This is obvious if $f(A, b, c) = \infty$. If $f(A, b, c) < \infty$, then the linear programming problem

$$\text{Min}\{c^T x \mid Ax = b, x \geq 0\} \tag{3.27}$$

is feasible. Let x be any feasible solution of it. Then, according to Theorem 2.13, x is also a nonnegative solution of the system (3.25); hence $\varphi \leq \underline{c}^T x \leq c^T x$, which implies (3.26). Thus (3.26) holds for each $A \in \mathbf{A}$, $b \in \mathbf{b}$, $c \in \mathbf{c}$, which means that $\varphi \leq \underline{f}$. Hence, from (a.1) and (a.2) we obtain $\underline{f} = \varphi$, which is (3.23).
   (b) To prove (3.24), put

$$\overline{\varphi} = \sup_{y \in Y_m} f(A_{ye}, b_y, \overline{c}).$$

(b.1) Since $A_{ye} \in \mathbf{A}$, $b_y \in \mathbf{b}$ for each $y \in Y_m$ and $\bar{c} \in \mathbf{c}$, we immediately obtain that

$$\overline{\varphi} \leq \sup\{f(A, b, c) \mid A \in \mathbf{A}, b \in \mathbf{b}, c \in \mathbf{c}\} = \overline{f}.$$

(b.2) Finally we prove $\overline{f} \leq \overline{\varphi}$ by showing that

$$f(A, b, c) \leq \overline{\varphi} \tag{3.28}$$

holds for each $A \in \mathbf{A}$, $b \in \mathbf{b}$, $c \in \mathbf{c}$. This is obvious if $f(A, b, c) = -\infty$. If $f(A, b, c) = \infty$, then the linear programming problem (3.27) is infeasible; hence the system $\mathbf{A}x = \mathbf{b}$ is not strongly feasible, which in view of Theorem 2.17 means that a system $A_{ye}x = b_y$ is not feasible for some $y \in Y_m$, so that $f(A_{ye}, b_y, \bar{c}) = \infty$; hence $\overline{\varphi} = \infty$ and (3.28) holds. Thus we are left with the case of $f(A, b, c)$ finite. Then by the duality theorem the dual problem to (3.27)

$$\text{Max}\{b^T p \mid A^T p \leq c\}$$

has an optimal solution p^* and $f(A, b, c) = b^T p^*$ holds. Put $y = \text{sgn}\, p^*$; then $y \in Y_m$ and $|p^*| = T_y p^*$. Consider the linear programming problem

$$\text{Min}\{\bar{c}^T x \mid A_{ye}x = b_y, x \geq 0\} \tag{3.29}$$

and its dual problem

$$\text{Max}\{b_y^T p \mid A_{ye}^T p \leq \bar{c}\}. \tag{3.30}$$

The dual problem (3.30) is solvable because p^* solves $A_{ye}^T p \leq \bar{c}$: in fact, since $|(A - A_c)^T p^*| \leq \Delta^T |p^*|$, we have

$$A_{ye}^T p^* = (A_c - T_y \Delta)^T p^* = (A_c^T - \Delta^T T_y) p^* = A_c^T p^* - \Delta^T |p^*|$$
$$\leq (A_c + A - A_c)^T p^* = A^T p^* \leq c \leq \bar{c}.$$

Now, if the primal problem (3.29) is infeasible, then $f(A_{ye}, b_y, \bar{c}) = \infty$; hence $\overline{\varphi} = \infty$ and (3.28) holds. If it is feasible, then $f(A, b, c) < \infty$ and $g(A, b, c) > -\infty$, and by the duality theorem the dual problem (3.30) has an optimal solution $\hat{p}$ satisfying $f(A_{ye}, b_y, \bar{c}) = b_y^T \hat{p}$; hence

$$f(A, b, c) = b^T p^* = (b_c + b - b_c)^T p^* \leq b_c^T p^* + \delta^T |p^*| = (b_c^T + \delta^T T_y) p^*$$
$$= (b_c + T_y \delta)^T p^* = b_y^T p^* \leq b_y^T \hat{p} = f(A_{ye}, b_y, \bar{c}) \leq \overline{\varphi},$$

which is (3.28). This proves that (3.28) holds for each $A \in \mathbf{A}$, $b \in \mathbf{b}$, $c \in \mathbf{c}$, implying $\overline{f} \leq \overline{\varphi}$. Hence, (b.1) and (b.2) together give $\overline{f} = \overline{\varphi}$, which proves (3.24). This completes the proof. $\qquad\square$

In Section 2.10 we presented an algorithm for checking feasibility of the systems

$$A_{ye}x = b_y$$

for all $y \in Y_m$. Since we are now facing a very close problem of solving

$$\text{Min}\{\bar{c}^T x \mid A_{ye} x = b_y, x \geq 0\}$$

for all $y \in Y_m$, we can adapt the previous algorithm for the current purpose. Let us reorder the equations of $Ax = b$ so that those containing at least one nondegenerate interval coefficient go first. Let q be the number of them, so that after reordering, the last $m - q$ equations consist only of real (i.e., noninterval) data. Then the following algorithm, where $z \in \mathbb{R}^q$ and $y \in \mathbb{R}^q$, does the job.

---

compute $\underline{f}$ by (3.23);
$z := 0; \ y := e;$
$A := \underline{A}; \ b := \overline{b}; \ \overline{f} := f(A, b, \overline{c});$
while $z \neq e$ & $\overline{f} < \infty$
  $k := \min\{i \mid z_i = 0\};$
  for $i := 1$ **to** $k - 1, \ z_i := 0;$ **end**
  $z_k := 1; \ y_k := -y_k;$
  if $y_k = 1$ **then** $A_{k\cdot} := \underline{A}_{k\cdot}; \ b_k := \overline{b}_k;$ **else** $A_{k\cdot} := \overline{A}_{k\cdot}; \ b_k := \underline{b}_k;$ **end**
  $\overline{f} := \max\{\overline{f}, f(A, b, \overline{c})\};$
end
% $[\underline{f}, \overline{f}]$ is the range of the optimal value.

Example 3.3. Let

$$c_c = (-1, -2, 3, 4)^T,$$

$$A_c = \begin{pmatrix} 5 & 6 & -7 & 8 \\ 10 & -11 & 12 & 13 \end{pmatrix}, \quad b_c = \begin{pmatrix} -9 \\ 14 \end{pmatrix}$$

(the pattern of the absolute values of coefficients is obvious). For each $\varepsilon > 0$ consider the interval data

$$\mathbf{A}_\varepsilon = [A_c - \varepsilon e e^T, A_c + \varepsilon e e^T], \quad \mathbf{b}_\varepsilon = [b_c - \varepsilon e, b_c + \varepsilon e], \quad \mathbf{c}_\varepsilon = [c_c - \varepsilon e, c_c + \varepsilon e].$$

Using the algorithm, we have computed $\underline{f}(\mathbf{A}_\varepsilon, \mathbf{b}_\varepsilon, \mathbf{c}_\varepsilon)$ and $\overline{f}(\mathbf{A}_\varepsilon, \mathbf{b}_\varepsilon, \mathbf{c}_\varepsilon)$ for $\varepsilon := 0.00, 0.01, \ldots, 0.24$ with MATLAB 6.0, where we employed the procedure QP.M for evaluating $f(A, b, c)$. The results, rounded to four decimal places, are summed up in the following table (the last column brings the values of $\overline{f}(\mathbf{A}_\varepsilon, \mathbf{b}_\varepsilon, \mathbf{c}_\varepsilon) - \underline{f}(\mathbf{A}_\varepsilon, \mathbf{b}_\varepsilon, \mathbf{c}_\varepsilon)$, denoted for short as $\overline{f} - \underline{f}$).

ε	$\underline{f}(\mathbf{A}_\varepsilon, \mathbf{b}_\varepsilon, \mathbf{c}_\varepsilon)$	$\overline{f}(\mathbf{A}_\varepsilon, \mathbf{b}_\varepsilon, \mathbf{c}_\varepsilon)$	$\overline{f} - \underline{f}$
0.00	5.0000	5.0000	0.0000
0.01	4.8085	5.2228	0.4143
0.02	4.6424	5.4847	0.8423
0.03	4.4971	5.7965	1.2994
0.04	4.3692	6.1735	1.8043
0.05	4.2559	6.6375	2.3816
0.06	4.1550	7.2217	3.0667
0.07	4.0647	7.9784	3.9137
0.08	3.9836	8.9955	5.0119
0.09	3.9104	10.4327	6.5223
0.10	3.8442	12.6143	8.7701
0.11	3.7841	16.3131	12.5290
0.12	3.7294	23.9388	20.2094
0.13	3.6796	48.7450	45.0654
0.14	3.6340	∞	∞
0.15	3.5923	∞	∞
0.16	3.5541	∞	∞
0.17	3.5189	∞	∞
0.18	3.4866	∞	∞
0.19	3.4569	∞	∞
0.20	3.4295	∞	∞
0.21	3.4043	∞	∞
0.22	3.3810	∞	∞
0.23	3.3396	∞	∞
0.24	$-\infty$	∞	∞

As we can see, all the linear programming problems in the family have optimal solutions for ε up to 0.13, infeasible problems appear from $\varepsilon = 0.14$ on and the family contains infeasible and unbounded problems (as well as those having optimal solutions) from $\varepsilon = 0.24$ on.

In the next two sections we study separately properties of the two bounds.

3.4 The lower bound

In this section we derive some consequences of the formula for the lower bound

$$\underline{f}(\mathbf{A}, \mathbf{b}, \mathbf{c}) = \inf\{\underline{c}^T x \mid \underline{A}x \le \overline{b}, \overline{A}x \ge \underline{b}, x \ge 0\}$$

in Theorem 3.2. If $\underline{f}(\mathbf{A}, \mathbf{b}, \mathbf{c}) = \infty$, then each problem in the family is infeasible. Let us first consider the case $\underline{f}(\mathbf{A}, \mathbf{b}, \mathbf{c}) = -\infty$.

Theorem 3.4. *If $\underline{f}(\mathbf{A}, \mathbf{b}, \mathbf{c}) = -\infty$, then there exists an $A_0 \in \mathbf{A}$ such that*

$$f(A_0, b, \underline{c}) \in \{-\infty, \infty\} \qquad (3.31)$$

holds for each $b \in \mathbf{b}$.

Comment. In other words, none of the problems

$$\text{Min}\{\underline{c}^T x \mid A_0 x = b, x \geq 0\}, \quad b \in \mathbf{b}$$

has an optimal solution.

Proof. If $\underline{f}(\mathbf{A}, \mathbf{b}, \mathbf{c}) = -\infty$, then by Theorem 3.2 the linear programming problem

$$\text{Min}\{\underline{c}^T x \mid \underline{A}x \leq \overline{b}, \overline{A}x \geq \underline{b}, x \geq 0\}$$

is unbounded and by the duality theorem its dual problem

$$\text{Max}\{\underline{b}^T p_1 - \overline{b}^T p_2 \mid \overline{A}^T p_1 - \underline{A}^T p_2 \leq \underline{c}, p_1 \geq 0, p_2 \geq 0\}$$

is infeasible, hence the system

$$\overline{A}^T p_1 - \underline{A}^T p_2 \leq \underline{c}$$

is infeasible and Theorem 2.7 assures existence of an x_0 that satisfies

$$\underline{A}x_0 \leq 0, \overline{A}x_0 \geq 0, x_0 \geq 0, \underline{c}^T x_0 < 0.$$

Then Theorem 2.13 gives that x_0 is a nonnegative weak solution of the system $[\underline{A}, \overline{A}]x = [0, 0]$; hence there exists a matrix $A_0 \in \mathbf{A}$ such that

$$A_0 x_0 = 0, \ x_0 \geq 0, \ \underline{c}^T x_0 < 0. \tag{3.32}$$

Now consider the problem

$$\text{Min}\{\underline{c}^T x \mid A_0 x = b, x \geq 0\} \tag{3.33}$$

for a $b \in \mathbf{b}$. If it is infeasible, then $f(A_0, b, \underline{c}) = \infty$. If it has a feasible solution x_1, then from (3.32) it follows that $x_1 + \alpha x_0$ is a feasible solution of (3.33) for each $\alpha \geq 0$ and

$$\lim_{\alpha \to \infty} \underline{c}^T (x_1 + \alpha x_0) = \lim_{\alpha \to \infty} (\underline{c}^T x_1 + \alpha \underline{c}^T x_0) = -\infty$$

due to (3.32); hence the problem (3.33) is unbounded and $f(A_0, b, \underline{c}) = -\infty$. Thus for each $b \in \mathbf{b}$ we have (3.31), which concludes the proof.   □

However, in the case of $\underline{f}(\mathbf{A}, \mathbf{b}, \mathbf{c}) = -\infty$ the family need *not* contain an unbounded problem.

Example 3.5. Let

$$\mathbf{A} = [0, 1], \quad \mathbf{b} = [1, 1], \quad \mathbf{c} = [-1, -1]$$

(i.e., $m = n = 1$). Then each problem in the family is of the form

$$\text{Min}\{-x \mid ax = 1, x \geq 0\},$$

it is infeasible for $a = 0$ and its optimal value is equal to $-1/a$ for $a \in (0, 1]$; hence $\underline{f}(\mathbf{A}, \mathbf{b}, \mathbf{c}) = -\infty$ but no problem in the family is unbounded.

If the lower bound $\underline{f}(\mathbf{A}, \mathbf{b}, \mathbf{c})$ is finite, then it can be expected that it is attained as the optimal value of some problem in the family. The following theorem shows a constructive way to find the data of such a problem.

Theorem 3.6. *Let $\underline{f}(\mathbf{A}, \mathbf{b}, \mathbf{c})$ be finite and let x^* be an optimal solution of the problem*

$$\text{Min}\{\underline{c}^T x \mid \underline{A}x \leq \overline{b}, \overline{A}x \geq \underline{b}, x \geq 0\}. \tag{3.34}$$

Then

$$\underline{f}(\mathbf{A}, \mathbf{b}, \mathbf{c}) = f(A_c - T_y \Delta, b_c + T_y \delta, \underline{c}), \tag{3.35}$$

where

$$y_i = \begin{cases} \frac{(A_c x^* - b_c)_i}{(\Delta x^* + \delta)_i} & \text{if } (\Delta x^* + \delta)_i > 0, \\ 1 & \text{if } (\Delta x^* + \delta)_i = 0 \end{cases} \qquad (i = 1, \ldots, m). \tag{3.36}$$

Proof. If $\underline{f}(\mathbf{A}, \mathbf{b}, \mathbf{c})$ is finite, then according to (3.23) it is equal to the optimal value of the problem (3.34); hence $\underline{f}(\mathbf{A}, \mathbf{b}, \mathbf{c}) = \underline{c}^T x^*$, where x^* is an arbitrary optimal solution of (3.34). Hence x^* satisfies

$$\underline{A}x^* \leq \overline{b}, \ \overline{A}x^* \geq \underline{b}, \ x^* \geq 0,$$

which can be equivalently written as

$$|A_c x^* - b_c| \leq \Delta x^* + \delta, \qquad x^* \geq 0. \tag{3.37}$$

Now Proposition 2.10 gives that $(A_c - T_y \Delta)x^* = b_c + T_y \delta$, where y is given by (3.36). Then $|y| \leq e$ because of (3.37); hence $A_c - T_y \Delta \in \mathbf{A}$ and $b_c + T_y \delta \in \mathbf{b}$, and we have

$$\underline{c}^T x^* = \underline{f}(\mathbf{A}, \mathbf{b}, \mathbf{c}) \leq f(A_c - T_y \Delta, b_c + T_y \delta, \underline{c}) \leq \underline{c}^T x^*,$$

which proves (3.35).   □

Notice that the vector y defined by (3.36) satisfies $y \notin Y_m$ in general (this is why we wrote $A_c - T_y \Delta$, $b_c + T_y \delta$ instead of A_{ye}, b_y in (3.35) because A_{ye}, b_y are defined for $y \in Y_m$ only; see Section 2.5). But y can be enforced to belong to Y_m under an additional assumption.

Theorem 3.7. *Let the problem*

$$\text{Max}\{\underline{b}^T p_1 - \overline{b}^T p_2 \mid \overline{A}^T p_1 - \underline{A}^T p_2 \leq \underline{c}, p_1 \geq 0, p_2 \geq 0\} \tag{3.38}$$

have an optimal solution p_1^*, p_2^* satisfying

$$p_1^* + p_2^* > 0. \tag{3.39}$$

Then

$$\underline{f}(\mathbf{A}, \mathbf{b}, \mathbf{c}) = f(A_{ye}, b_y, \underline{c}), \tag{3.40}$$

where

$$y_i = \begin{cases} 1 & \text{if } (p_2^*)_i > 0, \\ -1 & \text{if } (p_2^*)_i = 0 \end{cases} \qquad (i = 1, \ldots, m). \tag{3.41}$$

Proof. If the problem (3.38) has an optimal solution p_1^*, p_2^*, then its primal problem

$$\text{Min}\{\underline{c}^T x \mid \underline{A}x \leq \overline{b}, \overline{A}x \geq \underline{b}, x \geq 0\}$$

has an optimal solution x^*, $\underline{f}(\mathbf{A}, \mathbf{b}, \mathbf{c}) = \underline{c}^T x^*$ holds by Theorem 3.2 and the complementary slackness conditions of linear programming [31] give

$$p_1^{*T}(\overline{A}x^* - \underline{b}) = p_2^{*T}(\overline{b} - \underline{A}x^*) = 0.$$

Since all four vectors p_1^*, $\overline{A}x^* - \underline{b}$, p_2^* and $\overline{b} - \underline{A}x^*$ are nonnegative, it must be

$$(p_1^*)_i(\overline{A}x^* - \underline{b})_i = (p_2^*)_i(\overline{b} - \underline{A}x^*)_i = 0 \qquad (3.42)$$

for $i = 1, \ldots, m$. Now, if $(p_2^*)_i > 0$, then (3.42) gives $(\underline{A}x^*)_i = \overline{b}_i$; if $(p_2^*)_i = 0$, then $(p_1^*)_i > 0$ by (3.39) and (3.42) implies $(\overline{A}x^*)_i = \underline{b}_i$ $(i = 1, \ldots, m)$. Hence for the vector y defined by (3.41) we have $y \in Y_m$ and $A_{ye}x^* = b_y$, where $A_{ye} \in \mathbf{A}$ and $b_y \in \mathbf{b}$. Then

$$f(A_{ye}, b_y, \underline{c}) \leq \underline{c}^T x^* = \underline{f}(\mathbf{A}, \mathbf{b}, \mathbf{c}) \leq f(A_{ye}, b_y, \underline{c}),$$

which gives (3.40).   □

Finally we prove a kind of duality theorem for $\underline{f}(\mathbf{A}, \mathbf{b}, \mathbf{c})$ which shows that this value, if finite, can be reached via optimization over strong solutions of $A^T p \leq \mathbf{c}$ only (see Theorems 2.24 and 2.25).

Theorem 3.8. *If $\underline{f}(\mathbf{A}, \mathbf{b}, \mathbf{c})$ is finite, then*

$$\underline{f}(\mathbf{A}, \mathbf{b}, \mathbf{c}) = \max\{\min_{b \in \mathbf{b}} b^T p \mid A^T p \leq c \text{ for each } A \in \mathbf{A}, c \in \mathbf{c}\}. \qquad (3.43)$$

Proof. The proof consists of three steps.

(a) First we prove a technical result: for each $p \in \mathbb{R}^m$ there holds

$$\overline{A}^T p^+ - \underline{A}^T p^- = A_{ye}^T p, \qquad (3.44)$$

$$\underline{b}^T p^+ - \overline{b}^T p^- = b_y^T p = \min_{b \in \mathbf{b}} b^T p, \qquad (3.45)$$

where $y = -\text{sgn}\, p$. Indeed, since $|p| = -T_y p$, we have

$$\begin{aligned}
\overline{A}^T p^+ - \underline{A}^T p^- &= (A_c + \Delta)^T p^+ - (A_c - \Delta)^T p^- \\
&= A_c^T(p^+ - p^-) + \Delta^T(p^+ + p^-) \\
&= A_c^T p + \Delta^T |p| = A_c^T p - \Delta^T T_y p \\
&= (A_c - T_y \Delta)^T p = A_{ye}^T p,
\end{aligned}$$

which is (3.44), and

$$\begin{aligned}
b^T p = b^T(p^+ - p^-) &\geq \underline{b}^T p^+ - \overline{b}^T p^- = b_c^T(p^+ - p^-) - \delta^T(p^+ + p^-) \\
&= b_c^T p - \delta^T |p| = b_c^T p + \delta^T T_y p = (b_c + T_y \delta)^T p = b_y^T p
\end{aligned}$$

for each $b \in \mathbf{b}$; hence

$$\min_{b \in \mathbf{b}} b^T p \geq \underline{b}^T p^+ - \overline{b}^T p^- = b_y^T p \geq \min_{b \in \mathbf{b}} b^T p,$$

which is (3.45).

(b) If $\underline{f}(\mathbf{A}, \mathbf{b}, \mathbf{c})$ is finite, then by Theorem 3.2 and by the duality theorem we have

$$\underline{f}(\mathbf{A}, \mathbf{b}, \mathbf{c}) = \max\{\underline{b}^T p_1 - \overline{b}^T p_2 \mid \overline{A}^T p_1 - \underline{A}^T p_2 \leq \underline{c}, p_1 \geq 0, p_2 \geq 0\}. \quad (3.46)$$

Let p satisfy $A^T p \leq c$ for each $A \in \mathbf{A}$, $c \in \mathbf{c}$. Then in particular $A_{ye}^T p \leq \underline{c}$, where $y = -\operatorname{sgn} p$, and (3.44) gives $\overline{A}^T p^+ - \underline{A}^T p^- = A_{ye}^T p \leq \underline{c}$; hence from (3.45), (3.46) we obtain

$$\min_{b \in \mathbf{b}} b^T p = \underline{b}^T p^+ - \overline{b}^T p^-$$

$$\leq \max\{\underline{b}^T p_1 - \overline{b}^T p_2 \mid \overline{A}^T p_1 - \underline{A}^T p_2 \leq \underline{c}, p_1 \geq 0, p_2 \geq 0\}$$

$$= \underline{f}(\mathbf{A}, \mathbf{b}, \mathbf{c})$$

and consequently

$$\sup\{\min_{b \in \mathbf{b}} b^T p \mid A^T p \leq c \text{ for each } A \in \mathbf{A}, c \in \mathbf{c}\} \leq \underline{f}(\mathbf{A}, \mathbf{b}, \mathbf{c}). \quad (3.47)$$

(c) To prove that equality holds in (3.47), take any optimal solution p_1^*, p_2^* of the problem

$$\text{Max}\{\underline{b}^T p_1 - \overline{b}^T p_2 \mid \overline{A}^T p_1 - \underline{A}^T p_2 \leq \underline{c}, p_1 \geq 0, p_2 \geq 0\} \quad (3.48)$$

and put $\hat{p}_1 = p_1^* - d$, $\hat{p}_2 = p_2^* - d$, where $d = \min\{p_1^*, p_2^*\}$. Then $d \geq 0$, $\hat{p}_1 \geq 0$, $\hat{p}_2 \geq 0$ and $\hat{p}_1^T \hat{p}_2 = 0$. We show that $\hat{p}_1$, $\hat{p}_2$ is again an optimal solution of (3.48). In fact,

$$\overline{A}^T \hat{p}_1 - \underline{A}^T \hat{p}_2 = \overline{A}^T (p_1^* - d) - \underline{A}^T (p_2^* - d) = \overline{A}^T p_1^* - \underline{A}^T p_2^* - (\overline{A} - \underline{A})^T d$$

$$\leq \overline{A}^T p_1^* - \underline{A}^T p_2^* \leq \underline{c} \quad (3.49)$$

since $(\overline{A} - \underline{A})^T d \geq 0$, and

$$\underline{b}^T \hat{p}_1 - \overline{b}^T \hat{p}_2 = \underline{b}^T (p_1^* - d) - \overline{b}^T (p_2^* - d) = \underline{b}^T p_1^* - \overline{b}^T p_2^* + (\overline{b} - \underline{b})^T d$$

$$\geq \underline{b}^T p_1^* - \overline{b}^T p_2^*$$

since $(\overline{b} - \underline{b})^T d \geq 0$; hence it must be

$$\underline{b}^T \hat{p}_1 - \overline{b}^T \hat{p}_2 = \underline{b}^T p_1^* - \overline{b}^T p_2^* \quad (3.50)$$

and $\hat{p}_1$, $\hat{p}_2$ is an optimal solution of (3.48). Put $p = \hat{p}_1 - \hat{p}_2$. Since $\hat{p}_1^T \hat{p}_2 = 0$, it follows that $p^+ = \hat{p}_1$, $p^- = \hat{p}_2$; hence for each $A \in \mathbf{A}$, $c \in \mathbf{c}$ we have

$$A^T p = A^T(\hat{p}_1 - \hat{p}_2) \le \overline{A}^T \hat{p}_1 - \underline{A}^T \hat{p}_2 \le \underline{c} \le c$$

by (3.49), and

$$\min_{b \in \mathbf{b}} b^T p = \underline{b}^T p^+ - \overline{b}^T p^- = \underline{b}^T \hat{p}_1 - \overline{b}^T \hat{p}_2 = \underline{b}^T p_1^* - \overline{b}^T p_2^* = \underline{f}(\mathbf{A}, \mathbf{b}, \mathbf{c})$$

by (3.45), (3.50) and (3.46); hence the value $\underline{f}(\mathbf{A}, \mathbf{b}, \mathbf{c})$ is attained in (3.47) and (3.43) holds.   □

3.5 The upper bound

The formula

$$\overline{f}(\mathbf{A}, \mathbf{b}, \mathbf{c}) = \sup_{y \in Y_m} f(A_{ye}, b_y, \overline{c})$$

of Theorem 3.2 requires solving up to 2^m linear programs. In this section we show that the upper bound is closely connected with the optimal value of the nonlinear program

$$\text{maximize } b_c^T p + \delta^T |p| \tag{3.51}$$

s.t.

$$A_c^T p - \Delta^T |p| \le \overline{c}. \tag{3.52}$$

Let us denote

$$\overline{\varphi}(\mathbf{A}, \mathbf{b}, \mathbf{c}) = \sup\{b_c^T p + \delta^T |p| \mid A_c^T p - \Delta^T |p| \le \overline{c}\}. \tag{3.53}$$

We consider separately the cases of $\overline{\varphi}(\mathbf{A}, \mathbf{b}, \mathbf{c}) = -\infty$, $\overline{\varphi}(\mathbf{A}, \mathbf{b}, \mathbf{c}) = \infty$ and $\overline{\varphi}(\mathbf{A}, \mathbf{b}, \mathbf{c})$ finite.

Proposition 3.9. *If* $\overline{\varphi}(\mathbf{A}, \mathbf{b}, \mathbf{c}) = -\infty$, *then* $\overline{f}(\mathbf{A}, \mathbf{b}, \mathbf{c}) \in \{-\infty, \infty\}$.

Proof. Assume that $A^T p \le c$ has a solution for some $A \in \mathbf{A}$, $c \in \mathbf{c}$. Then $A_c^T p - \Delta^T |p| \le A^T p \le c \le \overline{c}$; hence p solves (3.52) (cf. Theorem 2.19), which implies that $\overline{\varphi}(\mathbf{A}, \mathbf{b}, \mathbf{c}) > -\infty$, a contradiction. Hence for each $A \in \mathbf{A}$, $b \in \mathbf{b}$, $c \in \mathbf{c}$ the dual problem

$$\text{Max}\{b^T p \mid A^T p \le c\}$$

is unsolvable, which means that each primal problem

$$\text{Min}\{c^T x \mid Ax = b, x \ge 0\}$$

is either infeasible or unbounded, so that $f(A, b, c) \in \{-\infty, \infty\}$ for each $A \in \mathbf{A}$, $b \in \mathbf{b}$, $c \in \mathbf{c}$ and consequently $\overline{f}(\mathbf{A}, \mathbf{b}, \mathbf{c}) \in \{-\infty, \infty\}$.   □

Proposition 3.10. *If* $\overline{\varphi}(\mathbf{A}, \mathbf{b}, \mathbf{c}) = \infty$, *then* $\overline{f}(\mathbf{A}, \mathbf{b}, \mathbf{c}) = \infty$.

Proof. If $\overline{\varphi}(\mathbf{A}, \mathbf{b}, \mathbf{c}) = \sup\{b_c^T p + \delta^T |p| \mid A_c^T p - \Delta^T |p| \leq \overline{c}\} = \infty$, then for each positive integer k there exists a $p_k \in \mathbb{R}^m$ such that

$$A_c^T p_k - \Delta^T |p_k| \leq \overline{c} \tag{3.54}$$

and

$$b_c^T p_k + \delta^T |p_k| \geq k. \tag{3.55}$$

For each $k = 1, 2, \ldots$ put $y_k = \operatorname{sgn} p_k$. Since $y_k \in Y_m$ for each k and Y_m is finite, the sequence $\{y_k\}_{k=1}^{\infty}$ must contain a member that appears there infinitely many times; i.e., there exists a subsequence $\{k_j\}_{j=1}^{\infty}$ and a $y \in Y_m$ such that $\operatorname{sgn} p_{k_j} = y$ for each j. Then from (3.54), (3.55) we have

$$A_{ye}^T p_{k_j} = A_c^T p_{k_j} - \Delta^T T_y p_{k_j} = A_c^T p_{k_j} - \Delta^T |p_{k_j}| \leq \overline{c},$$

$$b_y^T p_{k_j} = b_c^T p_{k_j} + \delta^T T_y p_{k_j} = b_c^T p_{k_j} + \delta^T |p_{k_j}| \geq k_j$$

for $j = 1, 2, \ldots$. Hence the problem

$$\operatorname{Max}\{b_y^T p \mid A_{ye}^T p \leq \overline{c}\}$$

is unbounded and by the duality theorem the respective primal problem

$$\operatorname{Min}\{\overline{c}^T x \mid A_{ye} x = b_y, x \geq 0\}$$

is infeasible; hence $f(A_{ye}, b_y, \overline{c}) = \infty$ and consequently $\overline{f}(\mathbf{A}, \mathbf{b}, \mathbf{c}) = \infty$. $\square$

Theorem 3.11. *If $\overline{\varphi}(\mathbf{A}, \mathbf{b}, \mathbf{c})$ is finite, then*

$$\overline{\varphi}(\mathbf{A}, \mathbf{b}, \mathbf{c}) = \max\{f(A, b, c) \mid f(A, b, c) < \infty, A \in \mathbf{A}, b \in \mathbf{b}, c \in \mathbf{c}\}. \tag{3.56}$$

Proof. (a) First we prove that if $f(A, b, c) < \infty$ for some $A \in \mathbf{A}$, $b \in \mathbf{b}$, $c \in \mathbf{c}$, then

$$f(A, b, c) \leq \overline{\varphi}(\mathbf{A}, \mathbf{b}, \mathbf{c}). \tag{3.57}$$

This is clearly the case if $f(A, b, c) = -\infty$. Thus let $f(A, b, c)$ be finite. Then $f(A, b, c) = b^T p^*$, where p^* is an optimal solution of the dual problem

$$\operatorname{Max}\{b^T p \mid A^T p \leq c\}.$$

Since

$$A_c^T p^* - \Delta^T |p^*| \leq A^T p^* \leq c \leq \overline{c},$$

we can see that p^* solves (3.52); hence

$$f(A, b, c) = b^T p^* \leq \sup\{b^T p \mid A_c^T p - \Delta^T |p| \leq \overline{c}\}$$
$$\leq \sup\{b_c^T p + \delta^T |p| \mid A_c^T p - \Delta^T |p| \leq \overline{c}\} = \overline{\varphi}(\mathbf{A}, \mathbf{b}, \mathbf{c}).$$

This proves (3.57) and hence also

$$\sup\{f(A,b,c) \mid f(A,b,c) < \infty, A \in \mathbf{A}, b \in \mathbf{b}, c \in \mathbf{c}\} \leq \overline{\varphi}(\mathbf{A},\mathbf{b},\mathbf{c}). \quad (3.58)$$

(b) To prove that the upper bound is attained in (3.58), we start from the fact that because of (3.53) for each positive integer k there exists a vector p_k satisfying

$$A_c^T p_k - \Delta^T |p_k| \leq \overline{c},$$

$$\overline{\varphi}(\mathbf{A},\mathbf{b},\mathbf{c}) - \tfrac{1}{k} < b_c^T p_k + \delta^T |p_k| \leq \overline{\varphi}(\mathbf{A},\mathbf{b},\mathbf{c}).$$

Arguing as in the proof of Proposition 3.10, we can assure existence of a $y \in Y_m$ satisfying $\operatorname{sgn} p_{k_j} = y$ for an infinite subsequence $\{k_j\}$. For each k_j we then have

$$A_{ye}^T p_{k_j} \leq \overline{c}, \quad (3.59)$$

$$\overline{\varphi}(\mathbf{A},\mathbf{b},\mathbf{c}) - \tfrac{1}{k_j} < b_y^T p_{k_j} \leq \overline{\varphi}(\mathbf{A},\mathbf{b},\mathbf{c}). \quad (3.60)$$

Now consider the problem

$$\operatorname{Max}\{b_y^T p \mid A_{ye}^T p \leq \overline{c}\}.$$

From (3.53) we have that its optimal value is bounded by $\overline{\varphi}(\mathbf{A},\mathbf{b},\mathbf{c})$, and (3.59), (3.60) show that this bound can be approximated with arbitrary accuracy by the value of the objective $b_y^T p$ over the solution set of $A_{ye}^T p \leq \overline{c}$. This gives, by the duality theorem,

$$\overline{\varphi}(\mathbf{A},\mathbf{b},\mathbf{c}) = \max\{b_y^T p \mid A_{ye}^T p \leq \overline{c}\} = f(A_{ye}, b_y, \overline{c});$$

hence the upper bound in (3.58) is attained and (3.56) holds.   □

Now we arrive at an important consequence that justifies introduction of the value $\overline{\varphi}(\mathbf{A},\mathbf{b},\mathbf{c})$.

Theorem 3.12. *If $\overline{f}(\mathbf{A},\mathbf{b},\mathbf{c})$ is finite, then*

$$\overline{f}(\mathbf{A},\mathbf{b},\mathbf{c}) = \overline{\varphi}(\mathbf{A},\mathbf{b},\mathbf{c}). \quad (3.61)$$

Proof. Since the possibilities $\overline{\varphi}(\mathbf{A},\mathbf{b},\mathbf{c}) = -\infty$ and $\overline{\varphi}(\mathbf{A},\mathbf{b},\mathbf{c}) = \infty$ are precluded by Propositions 3.9 and 3.10, $\overline{\varphi}(\mathbf{A},\mathbf{b},\mathbf{c})$ must be finite, and Theorem 3.11 gives

$$\overline{\varphi}(\mathbf{A},\mathbf{b},\mathbf{c}) = \max\{f(A,b,c) \mid f(A,b,c) < \infty, A \in \mathbf{A}, b \in \mathbf{b}, c \in \mathbf{c}\}$$
$$= \max\{f(A,b,c) \mid A \in \mathbf{A}, b \in \mathbf{b}, c \in \mathbf{c}\} = \overline{f}(\mathbf{A},\mathbf{b},\mathbf{c}).$$

□

Hence, if $\overline{f}(\mathbf{A},\mathbf{b},\mathbf{c})$ is finite, then it can be computed as the optimal value of a single nonlinear programming problem (3.51), (3.52) by nonlinear programming techniques. Moreover, the equality (3.61) yields a computable upper bound on $\overline{\varphi}(\mathbf{A},\mathbf{b},\mathbf{c})$. Let us recall that if A has linearly independent rows, then the matrix AA^T is nonsingular and there holds

$$(A^+)^T = (A^T)^+ = (AA^T)^{-1}A,$$

where A^+ is the Moore–Penrose inverse of A (Theorem 1.62, Corollary 1.63).

Theorem 3.13. *If A_c has linearly independent rows and if*

$$\varrho(\Delta|A_c^+|) < 1 \tag{3.62}$$

holds, then

$$\overline{\varphi}(\mathbf{A}, \mathbf{b}, \mathbf{c}) \leq \bar{c}^T|A_c^+|(I - \Delta|A_c^+|)^{-1}(|b_c| + \delta). \tag{3.63}$$

Proof. If the inequality (3.52) has no solution, then $\overline{\varphi}(\mathbf{A}, \mathbf{b}, \mathbf{c}) = -\infty$ and (3.63) holds. Thus let p be a solution to (3.52). Then we have

$$|p| = |(A_c A_c^T)^{-1} A_c A_c^T p| = |(A_c^+)^T A_c^T p| \leq |A_c^+|^T |A_c^T p|$$
$$\leq |A_c^+|^T (\Delta^T|p| + \bar{c}) \leq (\Delta|A_c^+|)^T |p| + |A_c^+|^T \bar{c};$$

hence

$$(I - \Delta|A_c^+|)^T |p| \leq |A_c^+|^T \bar{c}. \tag{3.64}$$

Because of (3.62) the matrix $I - \Delta|A_c^+|$ is nonnegatively invertible and premultiplying (3.64) by its transposed inverse gives

$$|p| \leq ((I - \Delta|A_c^+|)^{-1})^T |A_c^+|^T \bar{c}$$

and

$$b_c^T p + \delta^T |p| \leq (|b_c| + \delta)^T |p| \leq (|b_c| + \delta)^T ((I - \Delta|A_c^+|)^{-1})^T |A_c^+|^T \bar{c}$$
$$= \bar{c}^T |A_c^+|(I - \Delta|A_c^+|)^{-1}(|b_c| + \delta),$$

which yields (3.63). □

Theorem 3.12 can also be reformulated as a counterpart of Theorem 3.8 of Section 3.4.

Theorem 3.14. *If $\overline{f}(\mathbf{A}, \mathbf{b}, \mathbf{c})$ is finite, then*

$$\overline{f}(\mathbf{A}, \mathbf{b}, \mathbf{c}) = \max\{\max_{b \in \mathbf{b}} b^T p \mid A^T p \leq c \text{ for some } A \in \mathbf{A}, c \in \mathbf{c}\}. \tag{3.65}$$

Proof. By Gerlach's theorem 2.19, p satisfies (3.52) if and only if it is a weak solution of the system $[A_c^T - \Delta^T, A_c^T + \Delta^T]p \leq [\underline{c}, \bar{c}]$, i.e., if and only if it satisfies $A^T p \leq c$ for some $A \in \mathbf{A}$, $c \in \mathbf{c}$. Next, as in part (a) of the proof of Theorem 3.8 we can show that

$$b_c^T p + \delta^T |p| = b_y^T p = \max_{b \in \mathbf{b}} b^T p,$$

where $y = \operatorname{sgn} p$. The rest follows from Theorem 3.12. □

Observe the additional "duality" between quantifiers used in formulae (3.43) and (3.65): the optimization is performed over strong solutions in (3.43) and over weak solutions in (3.65).

Finally, we have this complexity result.

Theorem 3.15. *Computing the upper bound* $\overline{f}(\mathbf{A}, \mathbf{b}, \mathbf{c})$ *is NP-hard.*

Proof. Given a symmetric M-matrix A, consider the interval linear programming problem with $A_c = (A, -A)$, $\Delta = (0, 0)$, $b_c = 0$, $\delta = e$, $c_c = (e^T, e^T)^T$ and $\gamma = 0$. Then each primal problem in the family has the form

$$\text{Min}\{e^T x_1 + e^T x_2 \mid A(x_1 - x_2) = b, x_1 \geq 0, x_2 \geq 0\}$$

and its dual problem is of the form

$$\text{Max}\{b^T p \mid -e \leq Ap \leq e\}$$

(because A is symmetric by assumption). Each dual problem is solvable ($p = 0$ solves the system) and each solution p of $-e \leq Ap \leq e$ satisfies $|p| = |A^{-1} Ap| \leq |A^{-1}|e$; hence $|b^T p| \leq e^T |A^{-1}|e$, so that each dual problem has an optimal solution and thus also each primal problem has an optimal solution and the absolute value of its optimal value is bounded by $e^T |A^{-1}|e$. Hence $\overline{f}(\mathbf{A}, \mathbf{b}, \mathbf{c})$ is finite. Then by Theorem 3.12 we have

$$\overline{f}(\mathbf{A}, \mathbf{b}, \mathbf{c}) = \overline{\varphi}(\mathbf{A}, \mathbf{b}, \mathbf{c}) = \max\{e^T |p| \mid -e \leq Ap \leq e\};$$

hence

$$\overline{f}(\mathbf{A}, \mathbf{b}, \mathbf{c}) \geq 1 \qquad\qquad (3.66)$$

holds if and only if the system

$$-e \leq Ap \leq e,$$

$$e^T |p| \geq 1$$

has a solution. Since the latter problem is NP-complete by Theorem 2.3, the problem of deciding whether (3.66) holds is NP-hard; hence computing $\overline{f}(\mathbf{A}, \mathbf{b}, \mathbf{c})$ is NP-hard. $\qquad\square$

Summing up, we arrive at the following conclusion: computing the lower bound of the range of the optimal value $[\underline{f}(\mathbf{A}, \mathbf{b}, \mathbf{c}), \overline{f}(\mathbf{A}, \mathbf{b}, \mathbf{c})]$ can be performed in polynomial time, whereas computing the upper bound is NP-hard.

3.6 Finite range

In applications we are mostly interested in linear programming problems having optimal solutions. Therefore for problems with inexact data the case when all problems in the family have optimal solutions is of particular interest. Several equivalent conditions are listed in the following theorem.

Theorem 3.16. *For an interval linear programming problem with data* $\mathbf{A}$, $\mathbf{b}$, $\mathbf{c}$ *the following assertions are equivalent.*

(i)   *For each $A \in \mathbf{A}$, $b \in \mathbf{b}$, $c \in \mathbf{c}$ the problem*

$$\mathrm{Min}\{c^T x \mid Ax = b, x \geq 0\} \tag{3.67}$$

 has an optimal solution.
(ii) *Both $\underline{f}(\mathbf{A}, \mathbf{b}, \mathbf{c})$ and $\overline{f}(\mathbf{A}, \mathbf{b}, \mathbf{c})$ are finite.*
(iii) *Both $\underline{f}(\mathbf{A}, \mathbf{b}, \mathbf{c})$ and $\overline{\varphi}(\mathbf{A}, \mathbf{b}, \mathbf{c})$ are finite.*
(iv) *The system*

$$\overline{A}^T p_1 - \underline{A}^T p_2 \leq \underline{c} \tag{3.68}$$

 is feasible and $\overline{\varphi}(\mathbf{A}, \mathbf{b}, \mathbf{c})$ is finite.

In each case the range of the optimal value is given by

$$[\underline{f}(\mathbf{A}, \mathbf{b}, \mathbf{c}), \overline{\varphi}(\mathbf{A}, \mathbf{b}, \mathbf{c})].$$

Proof. We prove (i)$\Rightarrow$(ii)$\Rightarrow$(iii)$\Rightarrow$(iv)$\Rightarrow$(i).

(i)$\Rightarrow$(ii): Since each problem (3.67) has an optimal solution, it must be $\underline{f}(\mathbf{A}, \mathbf{b}, \mathbf{c}) < \infty$, and the possibility of $\underline{f}(\mathbf{A}, \mathbf{b}, \mathbf{c}) = -\infty$ is precluded by Theorem 3.4. Hence $\underline{f}(\mathbf{A}, \mathbf{b}, \mathbf{c})$ is finite, and Theorem 3.2 implies that $\overline{f}(\mathbf{A}, \mathbf{b}, \mathbf{c})$ is also finite.

(ii)$\Rightarrow$(iii) follows directly from Theorem 3.12.

(iii)$\Rightarrow$(iv): If $\underline{f}(\mathbf{A}, \mathbf{b}, \mathbf{c})$ is finite, then, as shown in part (b) of the proof of Theorem 3.8, equation (3.46), there holds

$$\underline{f}(\mathbf{A}, \mathbf{b}, \mathbf{c}) = \max\{\underline{b}^T p_1 - \overline{b}^T p_2 \mid \overline{A}^T p_1 - \underline{A}^T p_2 \leq \underline{c}, p_1 \geq 0, p_2 \geq 0\};$$

hence the system (3.68) is feasible.

(iv)$\Rightarrow$(i): Let the system (3.68) have a nonnegative solution p_1, p_2 and let $A \in \mathbf{A}$, $b \in \mathbf{b}$, $c \in \mathbf{c}$. Then we have

$$A^T(p_1 - p_2) \leq \overline{A}^T p_1 - \underline{A}^T p_2 \leq \underline{c} \leq c,$$

so that the dual problem to (3.67)

$$\mathrm{Max}\{b^T p \mid A^T p \leq c\} \tag{3.69}$$

is solvable, and for each solution p of $A^T p \leq c$ there holds

$$b^T p \leq \sup\{b^T p \mid A^T p \leq c\} \leq \sup\{b_c^T p + \delta^T |p| \mid A_c^T p - \Delta^T |p| \leq \overline{c}\}$$
$$= \overline{\varphi}(\mathbf{A}, \mathbf{b}, \mathbf{c}) < \infty;$$

hence the objective is bounded, so that the problem (3.69) has an optimal solution and by the duality theorem the problem (3.67) also has an optimal solution.

Since in all four cases $\overline{f}(\mathbf{A}, \mathbf{b}, \mathbf{c})$ is finite, we have $\overline{f}(\mathbf{A}, \mathbf{b}, \mathbf{c}) = \overline{\varphi}(\mathbf{A}, \mathbf{b}, \mathbf{c})$ by Theorem 3.12 and the range of the optimal value is equal to

$$[\underline{f}(\mathbf{A}, \mathbf{b}, \mathbf{c}), \overline{\varphi}(\mathbf{A}, \mathbf{b}, \mathbf{c})],$$

which concludes the proof. $\square$

Finally, we have this complexity result.

Theorem 3.17. *Checking whether each problem (3.67) with data satisfying $A \in \mathbf{A}$, $b \in \mathbf{b}$, $c \in \mathbf{c}$ has an optimal solution is NP-hard.*

Proof. Since a system $Ax = b$ is feasible if and only if the problem

$$\mathrm{Min}\{e^T x \mid Ax = b, x \geq 0\}$$

has an optimal solution, we have that a system of interval linear equations $\mathbf{A}x = \mathbf{b}$ is strongly feasible if and only if each problem (3.67) with data satisfying $A \in \mathbf{A}$, $b \in \mathbf{b}$, $c \in [e, e]$ has an optimal solution. Since the former problem is NP-hard by Theorem 2.18, the latter one is NP-hard as well. □

3.7 An algorithm for computing the range

Summing up Proposition 3.10 and Theorem 3.16, we can formulate the following alternative algorithm for computing the range which, in contrast to the algorithm of Section 3.3, requires solving two optimization problems only.

> compute the optimal value $\underline{f}$ of the problem
> $\quad \mathrm{Min}\{\underline{c}^T x \mid \underline{A}x \leq \overline{b}, \overline{A}x \geq \underline{b}, x \geq 0\}$;
> compute the optimal value $\overline{\varphi}$ of the problem
> $\quad \mathrm{Max}\{b_c^T p + \delta^T |p| \mid A_c^T p - \Delta^T |p| \leq \overline{c}\}$;
> **if** $\underline{f}$ is finite or $\overline{\varphi} = \infty$
> **then** $[\underline{f}, \overline{\varphi}]$ is the range of the optimal value
> **end**

3.8 Notes and references

In the last section we again give some additional notes and references to the material of the chapter.

Section 3.1. The duality theorem was published by Gale, Kuhn and Tucker [40] in 1951. The notion of it appeared earlier in an unpublished manuscript by J. von Neumann [192] which had evolved from his discussions with G. Dantzig in the autumn of 1947. Our formulation using functions $f(A, b, c)$ and $g(A, b, c)$ that may attain infinite values is atypical, but it allows us to formulate the duality theorem as well as two of its consequences in the form of a single equality (3.8).

Section 3.2. Although sensitivity analysis forms a standard part of linear programming textbooks, the interval linear programming problem was seemingly pioneered only in 1970 by Machost in his report [85]. His attempt to perform the simplex algorithm by replacing standard arithmetic operations by their interval arithmetic counterparts proved, however, to be ineffective;

moreover, the report contained some errors ([12, p. 8]). The first paper that handled the interval linear programming problem systematically was due to Krawczyk [79], followed by the state-of-the-art report by Beeck [12].

Section 3.3. As we have seen, Theorem 3.2 which gives formulae for computing $\underline{f}(\mathbf{A}, \mathbf{b}, \mathbf{c})$ and $\overline{f}(\mathbf{A}, \mathbf{b}, \mathbf{c})$ forms the cornerstone of our approach. The formula (3.23) for computing $\underline{f}(\mathbf{A}, \mathbf{b}, \mathbf{c})$, which is an easy consequence of the description of the set of nonnegative weak solutions of $\mathbf{A}x = \mathbf{b}$ by the system of inequalities (2.46), (2.47), appeared in [145]. The formula (3.24) for $\overline{f}(\mathbf{A}, \mathbf{b}, \mathbf{c})$ was proved in the report [150] (although for finite values only) and republished by Mráz in his survey paper [96]. The general treatment that allows for infinite values of $\overline{f}(\mathbf{A}, \mathbf{b}, \mathbf{c})$ presented here is new.

Section 3.4. The lower bound $\underline{f}(\mathbf{A}, \mathbf{b}, \mathbf{c})$ can be computed as the optimal value of the problem

$$\text{Min}\{\underline{c}^T x \mid \underline{A}x \leq \overline{b}, \overline{A}x \geq \underline{b}, x \geq 0\},$$

where the number of constraints is doubled compared to the original problem. But Theorems 3.6 and 3.7 suggest that one might also succeed with solving a problem of the original size with properly parameterized constraints. Mráz's report [91] is dedicated to this question; these and related results are summed up in his *habilitationsschrift* [94]. The "duality theorem" 3.8 for $\underline{f}(\mathbf{A}, \mathbf{b}, \mathbf{c})$ was published in [147] using a burdensome notation that obscured its actual contents (i.e., optimization over strong solutions of $\mathbf{A}^T p \leq \mathbf{c}$).

Section 3.5. While computing $\underline{f}(\mathbf{A}, \mathbf{b}, \mathbf{c})$ is easy, computation of $\overline{f}(\mathbf{A}, \mathbf{b}, \mathbf{c})$ is much more involved. Some partial results were achieved by Mráz in [92], [93], [95]. The treatment via $\overline{\varphi}(\mathbf{A}, \mathbf{b}, \mathbf{c})$, as presented in this section, is new. NP-hardness of computing $\overline{f}(\mathbf{A}, \mathbf{b}, \mathbf{c})$ was proved in technical report [162].

Section 3.6. The problem of finite range was first addressed in [149]; the conditions given in this section are easy consequences of our previous results. The NP-hardness result (Theorem 3.17) was proved in [164]. Based on the ideas outlined in this section, a "condition number" for linear programs was proposed in [155].

Section 3.7. The algorithm reduces the complicated formula (3.24) to solving only one nonlinear program. However, at the time this text was being written there was only limited computational experience at our disposal.

In our exposition we have left aside the difficult problem of determining (or bounding) the set of optimal solutions of all the linear programming problems contained in the family. A general treatment was done by Jansson [61], [62], [63], and computational aspects were studied by Jansson and Rump [65]. A special class is formed by so-called basis stable problems (where each problem in the family has a unique nondegenerate basic optimal solution with the same basis index set B) that were introduced by Krawczyk [79], characterized in [161], [150] and further studied by Koníčková [74], [75]. Basis stable problems are much easier to handle; but checking basis stability was proved to be NP-hard in an unpublished manuscript by Rohn.

Related works include Bauch et al. [9], Filipowski [38], Nedoma [98], Ramík [122], [127], Renegar [139], Vatolin [190] and Vera [191].

4

Linear programming with set coefficients

J. Nedoma and J. Ramík

4.1 Introduction

In this chapter we investigate a linear programming problem (LP problem) already formulated in Chapter 3. Here, we consider a family of linear programming problems

$$
\begin{aligned}
&\text{maximize } c^T x \\
&\text{subject to } Ax \leq b, \\
&\phantom{\text{subject to }} x \geq 0,
\end{aligned}
\tag{4.1}
$$

with data satisfying $c \in \mathbf{c} \subseteq \mathbb{R}^n$, $A \in \mathbf{A} \subseteq \mathbb{R}^{m \times n}$, $b \in \mathbf{b} \subseteq \mathbb{R}^m$, where $\mathbf{c}$, $\mathbf{A}$ and $\mathbf{b}$ are preselected sets. In comparison to Chapter 3, here $\mathbf{c}$, $\mathbf{A}$ and $\mathbf{b}$ are not necessarily (matrix or vector) intervals and the inequalities are considered in (4.1). The family of LP problems (4.1) is called the *linear programming problem with set coefficients (LPSC problem)*. In what follows, our interest is focused on the case where $\mathbf{c}$, $\mathbf{A}$ and $\mathbf{b}$ are either compact convex sets or, in particular, convex polytopes. We are interested primarily in the systems of inequalities in (4.1); later on we also deal with systems of equations. In the next section we start with the problem of duality of LPSC problems and later on we propose an algorithm, a generalized simplex method for solving such problems.

4.2 LP with set coefficients

Let $c^T = (c_1, \ldots, c_n)^T \in \mathbf{c} \subseteq \mathbb{R}^n$, $A = (a_{ij})_{i,j=1}^{m,n} \in \mathbf{A} \subseteq \mathbb{R}^{m \times n}$, $b^T = (b_1, \ldots, b_m)^T \in \mathbf{b} \subseteq \mathbb{R}^m$, where $\mathbf{c}$, $\mathbf{A}$ and $\mathbf{b}$ are preselected sets. Then LPSC problem (4.1) can be rewritten as follows,

$$\text{maximize } c_1 x_1 + \cdots + c_n x_n$$
$$\text{subject to } a_{i1}x_1 + \cdots + a_{in}x_n \leq b_i, \quad i \in \mathcal{M}, \tag{4.2}$$
$$x_j \geq 0, \quad j \in \mathcal{N}.$$

Here, $\mathcal{M} = \{1, 2, \ldots, m\}$ and $\mathcal{N} = \{1, 2, \ldots, n\}$.

We write the LPSC problem (4.1) briefly also as

$$\text{Max}\{c^T x | Ax \leq b, x \geq 0\} \tag{4.3}$$

with data satisfying $c \in \mathbf{c} \subseteq \mathbb{R}^n$, $A \in \mathbf{A} \subseteq \mathbb{R}^{m \times n}$, $b \in \mathbf{b} \subseteq \mathbb{R}^m$.

Corresponding to Chapter 2, we consider weak and strong feasibility of LPSC problem (4.1); moreover, we introduce a new concept: strict feasibility of the LPSC problem.

Let m be a positive integer and $\mathbf{U}, \mathbf{V}$ be subsets of $\mathbb{R}^m$. Define the following "inequality" relations,

$$\mathbf{U} \leq {}_1\mathbf{V} \text{ if } \forall\, u \in \mathbf{U}, \forall\, v \in \mathbf{V} : u \leq u,$$
$$\mathbf{U} \leq {}_2\mathbf{V} \text{ if } \exists\, u \in \mathbf{U}, \exists\, v \in \mathbf{V} : u \leq v.$$

For $t \in \{1, 2\}$ we define the following sets,

$$X_t^{\leq}(\mathbf{A}, \mathbf{b}) = \{x \in \mathbb{R}^n \mid \mathbf{A}x \leq_t \mathbf{b}, x \geq 0\}. \tag{4.4}$$

Here, $\mathbf{A}x = \{y \in \mathbb{R}^m | y = Ax, A \in \mathbf{A}\}$. The LPSC problem (4.2) is said to be *strictly feasible* if

$$X_1^{\leq}(\mathbf{A}, \mathbf{b}) \neq \emptyset.$$

If

$$X_2^{\leq}(\mathbf{A}, \mathbf{b}) \neq \emptyset,$$

we say that the LPSC problem (4.2) is *weakly feasible*. Moreover, we say that the LPSC problem (4.2) is *strongly feasible* if for each $A \in \mathbf{A}$ and each $b \in \mathbf{b}$ there exists $x \in \mathbb{R}^n, x \geq 0$, such that $Ax \leq b$.

In other words, given $\mathbf{A} \subseteq \mathbb{R}^{m \times n}$, $\mathbf{b} \subseteq \mathbb{R}^m$ an LPSC problem (4.2) is strictly feasible if there exists a vector $x \in \mathbb{R}^n, x \geq 0$, such that $Ax \leq b$ for all $A \in \mathbf{A}$ and all $b \in \mathbf{b}$. On the other hand, an LPSC problem (4.2) is weakly feasible if there exists a vector $x \in \mathbb{R}^n, x \geq 0$, and there exist some data $A \in \mathbf{A}$ and $b \in \mathbf{b}$ such that $Ax \leq b$. An LPSC problem (4.2) is strongly feasible if for each $A \in \mathbf{A}$ and each $b \in \mathbf{b}$ there exists a vector $x \in \mathbb{R}^n, x \geq 0$, such that $Ax \leq b$.

4.2.1 Strict, weak and strong feasibility

In this subsection, sufficient conditions for strong feasibility of the LPSC problem are derived. We show that in the special case of the interval LP problem strict feasibility and strong feasibility are equivalent.

By $\mathbb{R}^m_+$ we denote the usual nonnegative cone (orthant) in $\mathbb{R}^m$. Let $\mathbf{U}, \mathbf{V}$ be subsets of $\mathbb{R}^m$. By Max $\mathbf{U}$, Min $\mathbf{U}$, resp, Max $\mathbf{V}$, Min $\mathbf{V}$, we denote the corresponding sets of maximal and minimal elements of $\mathbf{U}$ and $\mathbf{V}$, resp. (with respect to the usual ordering $\leq$ in $\mathbb{R}^m$).

We also use the following notation

$$\text{Min}\mathbf{U} + \mathbb{R}^m_+ = \{u | u \in \mathbb{R}^m, u = u^- + u^+, u^- \in \text{Min}\mathbf{U}, u^+ \in \mathbb{R}^m_+\},$$

and similarly

$$\text{Max}\mathbf{U} + \mathbb{R}^m_+ = \{u | u \in \mathbb{R}^m, u = u^- + u^+, u^- \in \text{Max}\mathbf{U}, u^+ \in \mathbb{R}^m_+\}.$$

The following property is clear. If

$$\mathbf{U} \subseteq (\text{Min}\mathbf{U} + \mathbb{R}^m_+) \cap (\text{Max}\mathbf{U} - \mathbb{R}^m_+), \tag{4.5}$$
$$\mathbf{V} \subseteq (\text{Min } \mathbf{V} + \mathbb{R}^m_+) \cap (\text{Max } \mathbf{V} - \mathbb{R}^m_+),$$

then

$$\mathbf{U} \leq_1 \mathbf{V} \text{ if and only if } \forall\, x \in \text{Max } \mathbf{U},\ \forall\, y \in \text{Min } \mathbf{V} : x \leq y,$$
$$\mathbf{U} \leq_2 \mathbf{V} \text{ if and only if } \exists\, x \in \text{Min } \mathbf{U},\ \exists\, y \in \text{Max } \mathbf{V} : x \leq y.$$

Consider now the following LPSC problem

$$\begin{aligned} \text{maximize } & c^T x \\ \text{subject to } & Ax \leq b, \\ & x \geq 0, \end{aligned} \tag{4.6}$$

with data satisfying $A \in$ Max $\mathbf{A} \subseteq \mathbb{R}^{m \times n}$, $b \in$ Min $\mathbf{b} \subseteq \mathbb{R}^m$.

Proposition 4.1. *If the following conditions hold*

$$\mathbf{A} \subseteq (\text{Min } \mathbf{A} + \mathbb{R}^{m \times n}_+) \cap (\text{Max } \mathbf{A} - \mathbb{R}^{m \times n}_+), \tag{4.7}$$

$$\mathbf{b} \subseteq (\text{Min } \mathbf{b} + \mathbb{R}^m_+) \cap (\text{Max } \mathbf{b} - \mathbb{R}^m_+), \tag{4.8}$$

then

(i) $X^{\leq}_1(\mathbf{A}, \mathbf{b}) = \{x \in \mathbb{R}^n \mid \forall\, A \in \text{Max } \mathbf{A}, \forall\, b \in \text{Min } \mathbf{b} : Ax \leq b, x \geq 0\}$.
(ii) $X^{\leq}_2(\mathbf{A}, \mathbf{b}) = \{x \in \mathbb{R}^n \mid \exists A \in \text{Min } \mathbf{A}, \exists b \in \text{Max } \mathbf{b} : Ax \leq b, x \geq 0\}$.
(iii) Strong feasibility of problem (4.6) is equivalent to strong feasibility of (4.1).

Proof. (i) Let $R_1 = \{x \in \mathbb{R}^n \mid \forall\, A \in \text{Max } \mathbf{A}, \forall\, b \in \text{Min } \mathbf{b} : Ax \leq b, x \geq 0\}$. It is evident that $X^{\leq}_1(\mathbf{A}, \mathbf{b}) \subseteq R_1$. To prove the opposite inclusion, suppose $x \in R_1$, and take arbitrary $A \in \mathbf{A}$, $b \in \mathbf{b}$. By (4.7), (4.8) there are $A' \in$ Max $\mathbf{A}$, $b' \in$ Min $\mathbf{b}$, with $A \leq A', b' \leq b$. As $x \in R_1$, we have $A'x \leq b'$; consequently, $Ax \leq A'x \leq b' \leq b$, which means that $x \in X^{\leq}_1(\mathbf{A}, \mathbf{b})$.

(ii) Set $R_2 = \{x \in \mathbb{R}^n | \exists A \in \text{Min } \mathbf{A}, \exists b \in \text{Max } \mathbf{b} : Ax \leq b, x \geq 0\}$.
Apparently, $R_2 \subseteq X_{\bar{2}}^{\leq}(\mathbf{A}, \mathbf{b})$. In order to prove the opposite inclusion, suppose
that $x \in X_{\bar{2}}^{\leq}(\mathbf{A}, \mathbf{b})$; i.e., there exist $A \in \mathbf{A}, b \in \mathbf{b}$ with $Ax \leq b$. By (4.7),
(4.8) there are $A' \in \text{Min } \mathbf{A}$, $b' \in \text{Max } \mathbf{b}$, with $A' \leq A, b \leq b'$, which implies
$A'x \leq Ax \leq b \leq b'$; i.e., $x \in R_2$.

(iii) 1. Apparently, if (4.1) is strongly feasible, then (4.6) is strongly
feasible, too. 2. Suppose that (4.6) is strongly feasible. Choose arbitrarily
$A \in \mathbf{A}$, $b \in \mathbf{b}$; then by (4.7) and (4.8) there exist $A' \in \text{Max } \mathbf{A}$, $b' \in \text{Min }$
$\mathbf{b}$, such that $A \leq A'$, $b' \leq b$. As (4.6) is strongly feasible, there exists
$x' \geq 0$ with $A'x' \leq b'$. Thus, we obtain the following chain of consequences
$Ax' \leq A'x' \leq b' \leq b$, which means that x' is a nonnegative solution of (4.1);
i.e., the LPSC problem (4.1) is strongly feasible.   □

Let $\mathbf{A}$ be a matrix interval and $\mathbf{b}$ be a vector interval; i.e.,

$$\mathbf{A} = \{A \in \mathbb{R}^{m \times n} \mid \underline{A} \leq A \leq \bar{A}\}, \tag{4.9}$$
$$\mathbf{b} = \{b \in \mathbb{R}^m \mid \underline{b} \leq b \leq \bar{b}\},$$

where $\underline{A}$, $\bar{A}$, $\underline{b}$, $\bar{b}$ are given lower and upper bounds of the respective intervals. Then clearly, conditions (4.7), (4.8) are satisfied with $\text{Min } \mathbf{A} = \underline{A}$, Max
$\mathbf{A} = \bar{A}$, $\text{Min } \mathbf{b} = \underline{b}$, $\text{Max } \mathbf{b} = \bar{b}$. Then Proposition 4.1 can be reformulated as
follows.

$$X_{\bar{1}}^{\leq}(\mathbf{A}, \mathbf{b}) = \{x \in \mathbb{R}^n | \bar{A}x \leq \underline{b}, x \geq 0\}, \tag{4.10}$$

$$X_{\bar{2}}^{\leq}(\mathbf{A}, \mathbf{b}) = \{x \in \mathbb{R}^n | \underline{A}x \leq \bar{b}, x \geq 0\}. \tag{4.11}$$

Problem (4.1) is strongly feasible if and only if it is strictly feasible, or, if and
only if

$$X_{\bar{1}}^{\leq}(\mathbf{A}, \mathbf{b}) \neq \emptyset. \tag{4.12}$$

As was noticed in Chapter 3, checking strong feasibility of the LPSC problem (4.1) is important both from a theoretical and from a practical point of
view. Any strongly feasible problem (4.1) may be solved for any pair of the
data $A \in \mathbf{A}$, $b \in \mathbf{b}$. In the case of matrix or vector interval parameters $\mathbf{A}, \mathbf{b}$,
necessary and sufficient conditions for strong feasibility are relatively simple,
as in (4.12). In a general case, however, the situation is not so simple. Evidently, if problem (4.1) is strictly feasible then it is strongly feasible, but not
vice versa.

4.2.2 Objective function

In the LPSC problem (4.1) the vector coefficient c in objective function $z = c^T x$ is taken from the set $\mathbf{c} \subseteq \mathbb{R}^n$. Then we could solve the problem using one
of the following approaches.

$$z_0(x) = \sup\{c^T x \mid c \in \mathbf{c}\} \tag{4.13}$$

called the *"optimistic objective function"*,

$$z_1(x) = \inf\{c^T x \mid c \in \mathbf{c}\} \tag{4.14}$$

called the *"pessimistic objective function"*, or, a combination of the previous ones

$$z_s(x) = s z_0(x) + (1-s)z_1(x) \tag{4.15}$$

called the *"Hurwitz objective function"* where $s \in [0,1]$ is a *"parameter of optimism"*.

By a combination of objective function $z_s(x), s \in [0,1]$, with different types of feasible solution sets $X_t^{\leq}(\mathbf{A}, \mathbf{b})$, $t = 1, 2$, we obtain various optimization problems:

$(\mathbf{P}_{s,t})$

$$\begin{aligned} \text{maximize} \quad & z_s(x), \\ \text{subject to } & x \in X_t^{\leq}(\mathbf{A}, \mathbf{b}). \end{aligned} \tag{4.16}$$

The above problems have been dealt with by numerous authors; e.g., problem $(\mathbf{P}_{0,2})$ was formulated originally in [31], problem $(\mathbf{P}_{1,1})$ was analyzed by [188], [189] and some of the other problems were investigated in [119], [123], [98], and others.

4.3 Duality

In this section, we deal with the problem of duality in LPSC; see also [189]. Let us start with the definitions of primal LPSC problem (P) and dual LPSC problem (D). Let $\mathbf{c} \subseteq \mathbb{R}^n$, $\mathbf{A} \subseteq \mathbb{R}^{m \times n}$, $\mathbf{b} \subseteq \mathbb{R}^m$ be nonempty sets.

The *primal LPSC problem* (P) is a family of LP problems

$$\begin{aligned} \text{maximize} \quad & c^T x, \\ \text{subject to } & Ax \leq b, \\ & x \geq 0, \end{aligned} \tag{4.17}$$

with coefficients satisfying $c \in \mathbf{c}$, $A \in \mathbf{A}$, $b \in \mathbf{b}$.

The *dual LPSC problem* (D) is a family of LP problems

$$\begin{aligned} \text{minimize} \quad & b^T y, \\ \text{subject to } & A^T y \geq c, \\ & y \geq 0, \end{aligned} \tag{4.18}$$

with coefficients satisfying $c \in \mathbf{c}$, $A \in \mathbf{A}$, $b \in \mathbf{b}$.

As the pair of problems (P), (D) is not the usual primal–dual pair of LP problems, it is necessary to specify how duality will be understood, particularly, what are the solutions of (P) and (D).

We define the *set of feasible solutions of primal LPSC problem (P)* as $X_2^{\leq}(\mathbf{A}, \mathbf{b})$, originally defined by (4.4).

Here, for the sake of simplicity, we omit the subscript 2; then we have

$$X^{\leq}(\mathbf{A}, \mathbf{b}) = \{x \in \mathbb{R}^n \mid x \geq 0, \exists\, A \in \mathbf{A}, \exists\, b \in \mathbf{b} : Ax \leq b\}. \tag{4.19}$$

The *set of feasible solutions of the dual LPSC problem (D)*, denoted by $Y^{\geq}(\mathbf{A}, \mathbf{c})$, is defined as follows.

$$Y^{\geq}(\mathbf{A}, \mathbf{c}) = \{y \in \mathbb{R}^m \mid y \geq 0, \forall\, A \in \mathbf{A}, \forall\, c \in \mathbf{c} : A^T y \geq c\}. \tag{4.20}$$

Now, the *optimal value of the primal LPSC problem (P)*, denoted by $f(\mathbf{A}, \mathbf{b}, \mathbf{c})$, is defined as

$$f(\mathbf{A}, \mathbf{b}, \mathbf{c}) = \sup\{c^T x \mid x \in X^{\leq}(\mathbf{A}, \mathbf{b})\,,\, c \in \mathbf{c}\}, \tag{4.21}$$

and the *optimal value of the dual LPSC problem (D)*, denoted by $g(\mathbf{A}, \mathbf{b}, \mathbf{c})$, is defined as

$$g(\mathbf{A}, \mathbf{b}, \mathbf{c}) = \inf\{b^T y \mid y \in Y^{\geq}(\mathbf{A}, \mathbf{c})\,,\, b \in \mathbf{b}\}. \tag{4.22}$$

If we have

$$c^* \in \mathbf{c},\ x^* \in X^{\leq}(\mathbf{A}, \mathbf{b})\ \text{and}\ c^{*T} x^* = f(\mathbf{A}, \mathbf{b}, \mathbf{c}),$$

then x^* is said to be an *optimal solution of the primal LPSC problem (P)*.

If we have

$$b^* \in \mathbf{b},\ y^* \in Y^{\geq}(\mathbf{A}, \mathbf{c})\ \text{and}\ b^{*T} y^* = g(\mathbf{A}, \mathbf{b}, \mathbf{c}),$$

then y^* is called an *optimal solution of the dual LPSC problem (D)*.

In the next subsections, we investigate relations between problems (P) and (D) and derive weak and strong duality results.

4.3.1 Weak duality

We begin with the simpler version of the weak duality theorem for the case that $\mathbf{b}$ is a singleton.

Theorem 4.2. *Let $\mathbf{b} = \{b\}$; i.e., $\mathbf{b}$ is a singleton in $\mathbb{R}^m$. Then*

$$f(\mathbf{A}, \{b\}, \mathbf{c}) \leq g(\mathbf{A}, \{b\}, \mathbf{c}). \tag{4.23}$$

If there exist $c^* \in \mathbf{c}$, $x^* \in X^{\leq}(\mathbf{A}, \{b\})$, $y^* \in Y^{\geq}(\mathbf{A}, \mathbf{c})$ such that

$$c^{*T} x^* = b^T y^*, \tag{4.24}$$

then x^* is an optimal solution of the primal LPSC problem (P) and y^* is an optimal solution of the dual LPSC problem (D).

Proof. 1. If $X^{\le}(\mathbf{A}, \{b\}) = \emptyset$, or $Y^{\ge}(\mathbf{A}, \mathbf{c}) = \emptyset$, then (4.23) trivially holds.

2. Let $x \in X^{\le}(\mathbf{A}, \{b\})$, and choose arbitrarily $y \in Y^{\ge}(\mathbf{A}, \mathbf{c})$. Then there is $A \in \mathbf{A}$, such that $Ax \le b$. For all $c \in \mathbf{c}$ we have $y^T A \ge c^T$; i.e., $y^T A x \ge c^T x$. Moreover, $y^T A x \le b^T y$, consequently, $c^T x \le b^T y$ for all $c \in \mathbf{c}$, which means

$$f(\mathbf{A}, \{b\}, \mathbf{c}) \le b^T y.$$

Since y was arbitrary, we obtain the required result (4.23).

3. Apparently,

$$c^{*T} x^* \le f(\mathbf{A}, \{b\}, \mathbf{c}) \le g(\mathbf{A}, \{b\}, \mathbf{c}) \le b^T y^*,$$

and by (4.24) we obtain equations in the above chain of inequalities.   □

Corollary 4.3. *If* **b** *consists of more than one element, then*

$$f(\mathbf{A}, \mathbf{b}, \mathbf{c}) \le \sup\{g(\mathbf{A}, \{b\}, \mathbf{c}); b \in \mathbf{b}\}, \tag{4.25}$$

$$\inf\{f(\mathbf{A}, \{b\}, \mathbf{c}); b \in \mathbf{b}\} \le g(\mathbf{A}, \mathbf{b}, \mathbf{c}). \tag{4.26}$$

Proof. For each $b \in \mathbf{b}$ we obtain by (4.23) $f(\mathbf{A}, \{b\}, \mathbf{c}) \le g(\mathbf{A}, \{b\}, \mathbf{c})$. Applying supremum on both sides, we obtain (4.25); doing infimum, we get (4.26).   □

In general, the strict inequality " < " in (4.24) can occur, as the following example shows.

Let **A** *be a "matrix segment"*

$$\mathbf{A} = \left\{ A_\lambda \mid A_\lambda = \begin{pmatrix} 1 & -\lambda \\ \lambda & 1 \end{pmatrix}, 0 \le \lambda \le 1 \right\},$$

$\mathbf{c} = \{c\} = \{(2,3)^T\}$, $\mathbf{b} = \{b\} = \{(1,1)^T\}$. *The following statements are clearly true.*

(i) **A** *is regular, convex and compact in* $\mathbb{R}^{2 \times 2}$.
(ii) $f(\mathbf{A}, \{b\}, \{c\}) = 5 < 7 = g(\mathbf{A}, \{b\}, \{c\})$.

Here, the regularity of **A** *means that each matrix* $A \in \mathbf{A}$ *is nonsingular in the usual sense. Notice, that A is a square* 2×2 *matrix, and determinant* $\det(A) = 1 + \lambda^2$ *for all* $A \in \mathbf{A}$; *hence* **A** *is regular.*

The above example shows that sufficient conditions which eventually secure strong duality, i.e., equality in (4.24), will not be very simple. For instance, convexity, compactness and regularity of the set **A** is not sufficient for strong duality. In the next subsection we look for such sufficient (eventually, necessary and sufficient) conditions.

4.3.2 Strong duality

Before formulating the main result, some modifications of the original problem are useful.

Modification 1.
Using the reformulation of the system of constraints (4.1) by shifting the right- hand side to the left side and incorporating the set $\mathbf{b}$ into the new left side, we obtain the new right side which is a fixed vector; i.e.,

$$
\begin{aligned}
Ax - bx_{n+1} &\leq 0, \\
x_{n+1} &\leq 1, \\
-x_{n+1} &\leq -1, \\
x_j \geq 0, \ j = 1, 2, \dots, n, \ n+1.
\end{aligned} \tag{4.27}
$$

We obtain a new system of constraints (4.1) with $\mathbf{b} = \{b\} = \{(0, 1, -1)^T\}$.

Modification 2.
From now on, we consider the LPSC problem (4.1), in the standard form; i.e., we consider the LPSC problem with equalities,

$$
\text{Max}\{c^T x | Ax = b, x \geq 0\}, \tag{4.28}
$$

with coefficients satisfying $c \in \mathbf{c}$, $A \in \mathbf{A}$, $\mathbf{b} = \{(0, 1, -1)^T\}$. Transformation from the canonical form (with inequalities " $\leq$ ") to the standard form makes no difficulties and requires only additional variables supplemented to the individual rows on the left side of the constraints (4.1).

The dual couple of LPSC problems may be formulated after the above modifications as follows.

(P*)

$$
\begin{aligned}
\text{maximize} \quad & c^T x, \\
\text{subject to } & Ax = b, \\
& x \geq 0,
\end{aligned} \tag{4.29}
$$

with coefficients satisfying $c \in \mathbf{c}$, $A \in \mathbf{A}$.

(D*)

$$
\begin{aligned}
\text{minimize} \quad & b^T y, \\
\text{subject to } & A^T y \geq c,
\end{aligned} \tag{4.30}
$$

with the coefficients satisfying $c \in \mathbf{c}$, $A \in \mathbf{A}$.

Similarly to the previous subsection (see (4.19), (4.20)) we define the corresponding feasible solution sets for (P*) and (D*), respectively:

$$
X^=(\mathbf{A}, \{b\}) = \{x \in \mathbb{R}^n \mid x \geq 0, \ \exists A \in \mathbf{A} : Ax = b\}, \tag{4.31}
$$

$$Y^{\geq}(\mathbf{A}, \mathbf{c}) = \{y \in \mathbb{R}^m \mid \forall A \in \mathbf{A}, \ \forall c \in \mathbf{c} : A^T y \geq c\}, \qquad (4.32)$$

According to the above modifications, we can redefine the optimal values of the dual problems as follows.

$$f(\mathbf{A}, \{b\}, \mathbf{c}) = \sup\{c^T x \mid x \in X^{=}(\mathbf{A}, \{b\}), \ c \in \mathbf{c}\}, \qquad (4.33)$$

$$g(\mathbf{A}, \{b\}, \mathbf{c}) = \inf\{b^T y \mid y \in Y^{\geq}(\mathbf{A}, \mathbf{c})\}.$$

Now, we consider the following assumptions.

Assumption I: $m = n$; i.e., $\mathbf{A} \subseteq \mathbb{R}^{n \times n}$.

Assumption II: $\mathbf{A}$ is compact and regular.

Assumption III: $\mathbf{A} = (\mathbf{A}_1, \dots, \mathbf{A}_n)$, where $\mathbf{A}_1 \subseteq \mathbb{R}^m, \dots, \mathbf{A}_n \subseteq \mathbb{R}^m$; i.e., $\mathbf{A}$ is "columnwise separable" or, in other words, $A \in \mathbf{A}$,

$$A = \begin{pmatrix} a_{11} & \dots & a_{1n} \\ a_{21} & \dots & a_{2n} \\ \vdots & \dots & \vdots \\ a_{n1} & \dots & a_{nn} \end{pmatrix}, \text{ if } \begin{pmatrix} a_{11} \\ a_{21} \\ \vdots \\ a_{n1} \end{pmatrix} \in \mathbf{A}_1, \dots, \begin{pmatrix} a_{1n} \\ a_{2n} \\ \vdots \\ a_{nn} \end{pmatrix} \in \mathbf{A}_n.$$

Prior to proving the strong duality theorem, we need the following two lemmas, the first of which is based on the result of [98].

Lemma 4.4. *Let Assumptions I, II, III be satisfied; let $c \in \mathbb{R}^n$. Then there exist $y^* \in Y^{\geq}(\mathbf{A}, c)$ and $A^* \in \mathbf{A}$ such that*

$$A^{*T} y^* = c; \qquad (4.34)$$

i.e.,

$$A^* = (A_1^*, \dots, A_n^*) \quad \text{and} \quad A_j^* \in \mathbf{A}_j, \quad j = 1, \dots, n.$$

Moreover,

$$A_j^{*T} y^* = c_j, \quad \text{for } j = 1, \dots, n. \qquad (4.35)$$

Proof. The proof is carried out by induction on k, where

$$\mathbf{A} = \mathbf{A}^k = (\mathbf{A}_1, \mathbf{A}_2, \dots, \mathbf{A}_k, \{A_{k+1}\}, \dots, \{A_n\}), \qquad (4.36)$$

i.e. the first k columns of the set $\mathbf{A}$ are variable; the rest of the $n - k$ columns are definite vectors.

1. For $k = 0$, we have $\mathbf{A} = \{A\} = \{A\}^*$, A is regular and equation (4.34) or (4.35) has a single solution $y^* \in Y^{\geq}(A, c)$.

2. Assume that the lemma holds for $k - 1 \leq n - 1$ and consider

$$\mathbf{A} = \mathbf{A}^k = (\mathbf{A}_1, \mathbf{A}_2, \dots, \mathbf{A}_k, \{A_{k+1}\}, \dots, \{A_n\}). \qquad (4.37)$$

For an arbitrary $A_{k,0} \in \mathbf{A}_k$ there exists $y_0 \in Y^{\geq}(\mathbf{A}, \{c\})$ and $A_{j,0} \in \mathbf{A}_j$, $j = 1, 2, \ldots, k - 1$, satisfying

$$A_{j,0}^T y_0 = c_j, \text{ for } j = 1, \ldots, k, \text{ and } A_j^T y_0 = c_j, \text{ for } j = k+1, \ldots, n. \quad (4.38)$$

Define the sequences $A_{k,n} \in \mathbf{A}_k$, $y_n \in Y^{\geq}(\mathbf{A}, c)$ recursively as follows.

If $A_{k,n-1} \in \mathbf{A}_k$, then there exist $y_{n-1} \in Y^{\geq}(\mathbf{A}, c)$, and $A_{j,n-1} \in \mathbf{A}_j$, $j = 1, 2, \ldots, k - 1$ satisfying

$$A_{j,n-1}^T y_{n-1} = c_j, \text{ for } j = 1, \ldots, k, \text{ and } A_j^T y_{n-1} = c_j, \text{ for } j = k+1, \ldots, n. \quad (4.39)$$

Then $A_{k,n-1} \in \mathbf{A}_k$ is determined by

$$A_{k,n-1}^T y_{n-1} = \min\{A_k^T y_{n-1} | \ A_k \in \mathbf{A}_k\}. \quad (4.40)$$

The minimum in (4.40) exists, as $\mathbf{A}_k$ is compact. Let $\bar{A}_k$ be an accumulation point of the sequence $\{A_{k,n}\}$, without loss of generality suppose that $A_{k,n} \to \bar{A}_k$, as $n \to \infty$, apparently; $\bar{A}_k \in \mathbf{A}_k$. There exists a subsequence of $\{y_n\}$, $y_{n_\ell} \to \bar{y}$, as $\ell \to \infty$; $\bar{y} \in Y^{\geq}(\mathbf{A}, c)$. Since by (4.39) we have $A_{k,n_\ell}^T y_{n_\ell} = c_k$, as $\ell \to \infty$, we obtain $\bar{A}_k^T \bar{y} = c_k$. It remains to prove that $A_k^T \bar{y} \geq c_k$ for all $A_k \in \mathbf{A}_k$.

For any $A_k \in \mathbf{A}_k$, the inequalities

$$A_k^T y_{i-1} \geq A_{k,i-1}^T y_{i-1} = c_k, \quad i = 1, 2, \ldots$$

hold according to (4.40) and hence

$$A_k^T \bar{y} \geq \bar{A}_k^T \bar{y} = c_k.$$

$\square$

Lemma 4.5. *Let Assumption I and Assumption II be satisfied. Then*

$$X^{=}(\mathbf{A}, \{b\}) = \{x \in \mathbf{E}^n \ | \ x \geq 0, \exists A \in \mathbf{A} : Ax = b\}$$

is compact.

Proof. Let $x_n \in X^{=}(\mathbf{A}, b)$, $i = 1, 2, \ldots$; then there exist $A_n \in \mathbf{A}$ with $A_n x_n = b$. Since A_n is regular and $\mathbf{A}$ is compact, there are subsequences $A_{n_k} \to \bar{A} \in \mathbf{A}$, and $x_{n_k} \to \bar{x}$, $\bar{x} \geq 0$, for $k \to \infty$, and $\bar{A}\bar{x} = b$. Hence $\bar{x} \in X^{=}(\mathbf{A}, b)$.

$\square$

Theorem 4.6. *(Strong duality theorem)*
   Let Assumptions I, II, III be satisfied, let $\mathbf{c}$ *be compact and LPSC problem (4.29) be strongly feasible. Then*

$$f(\mathbf{A}, \{b\}, \mathbf{c}) = g(\mathbf{A}, \{b\}, \mathbf{c}); \quad (4.41)$$

moreover, there exist $c^* \in \mathbf{c}$, $x^* \in X^{=}(\mathbf{A}, \{b\})$, $y^* \in Y^{\geq}(\mathbf{A}, \{c^*\})$ *such that*

$$c^{*T} x^* = b^T y^*. \quad (4.42)$$

Proof. By Lemma 4.4. there exist $c^* \in \mathbf{c}$ and $x^* \in X^=(\mathbf{A}, \{b\})$ such that

$$c^{*T}x^* = \text{Max}\{c^Tx \mid x \in X^=(\mathbf{A}, \{b\}), c \in \mathbf{c}\} = f(\mathbf{A}, \{b\}, \mathbf{c}).$$

By Lemma 4.4 there exist $y^* \in Y^{\geq}(\mathbf{A}, \{c^*\})$ and $A^* \in \mathbf{A}$ with $A^{*T}y^* = c^*$. Using strong feasibility we obtain $x' \geq 0$, with $A^*x' = b$. Hence

$$x'^T A^{*T} y^* = c^{*T} x' = b^T y^*, \tag{4.43}$$

giving (4.42). Generally, we have the following chain of inequalities

$$c^{*T}x' \leq f(\mathbf{A}, \{b\}, \mathbf{c}) \leq g(\mathbf{A}, \{b\}, \mathbf{c}) \leq b^T y^*;$$

hence by (4.43) we obtain (4.41). $\qquad\square$

Theorem 4.7. *Let the assumptions of Lemma 4.4 be satisfied, let* $\mathbf{c} = \{c\}$ *and let* $A^* = (A_1^*, \ldots, A_n^*)$, $A_j^* \in \mathbf{A}_j$, $j = 1, \ldots, n$, *be from Lemma 4.4. Then the optimal solution of the LP problem:*

$$\text{Max}\{c^Tx \mid A^*x = b, \ x \geq 0\} \tag{4.44}$$

is the optimal solution of the LPSC problem (4.29), i.e., problem (P*).

Proof. Let x^* be an optimal solution of (4.44), and let $y^* \in Y^{\geq}(\mathbf{A}, \{c\})$ be from Lemma 4.4. Then
$$A^{*T}y^* = c;$$

hence

$$x^{*T} A^{*T} y^* = c^T x^* = b^T y^*,$$

and consequently, x^* is an optimal solution of (P*). $\qquad\square$

The last theorem gives a simple criterion of how to find an optimal solution of LPSC problem (4.29), on condition that the matrix A^* from Lemma 4.4 is known. The algorithmic problem now is how to find A^*, or, eventually, y^* ? This problem is dealt with in the next section.

4.4 Generalized simplex method

The aim of this section is to derive the generalized simplex method for solving LPSC problem (4.29); see [98]. In the previous section we have derived a criterion which is, however, applicable only to problems with a square matrix in the constraints (4.29). In the next subsection we propose an algorithm that "works" under Assumptions I, II, III and finds the dual optimal solution.

4.4.1 Finding the dual optimal solution: Algorithm

```
select y* ≠ 0;
 repeat
    y := y*;
    for j := 1 to n
        find a* ∈ A_j for which y^T a* = Min{y^T a | a ∈ A_j};
        A_j* := a*;
    end
        y* := (A*^T)^{-1}c;
 until y* = y
 % A*, y* is an optimal solution of (D*).
```

Theorem 4.8. *On the assumptions of the strong duality theorem, Algorithm 4.4.1 generates a sequence $\{A^n\}$ with the following property. If A^* is an accumulation point of $\{A^n\}$, then $y^* = (A^{*T})^{-1}c$ is an optimal solution of dual problem (4.30), i.e., of (D*).*

Proof. Let $A \in \mathbf{A}$; then the construction of the innermost optimization problem in Algorithm (4.4.1) assures that $A^T y^* \geq A^{*T} y^* = c$. Then by Theorem 4.6, we get $f(\mathbf{A}, \{b\}, \{c\}) = g(\mathbf{A}, \{b\}, \{c\})$. Hence y^* is an optimal solution of dual problem (D*). □

Let A_j be polyhedrons and $j = 1, \ldots, n$. Then the algorithm converges in a finite number of iterations.

Example 4.9. Consider the following LPSC problem,
(P*)

$$\text{maximize } 3x_1 + x_2,$$
$$\text{subject to } a_1 x_1 + 2x_2 = 3,$$
$$a_2 x_1 + x_2 = 4,$$
$$x_1, x_2 \geq 0,$$

where

$$\mathbf{A}_1 = \left\{ \mathbf{a} \in \mathbb{R}^2 \mid \mathbf{a} = \lambda \begin{pmatrix} 1 \\ 2 \end{pmatrix} + (1 - \lambda) \begin{pmatrix} 2 \\ 3 \end{pmatrix}, \ \lambda \in [0, 1] \right\},$$

$$\mathbf{A}_2 = \left\{ \begin{pmatrix} 2 \\ 1 \end{pmatrix} \right\};$$

i.e., $\mathbf{A} = (\mathbf{A}_1, \mathbf{A}_2) \subseteq \mathbb{R}^{2 \times 2}$. The coefficients in the first column of the matrix belong to the segment $\mathbf{A}_1$, whereas the second column is a fixed vector $\begin{pmatrix} 2 \\ 1 \end{pmatrix}$.

Moreover, $b = \begin{pmatrix} 3 \\ 4 \end{pmatrix}$ and $c = \begin{pmatrix} 3 \\ 1 \end{pmatrix}$. It is easy to verify that Assumptions I, II, III are satisfied. The dual problem can be formulated as follows.
(D*)

$$\text{minimize } 3y_1 + 4y_2,$$
$$\text{subject to } a_1y_1 + a_2y_2 \geq 3,$$
$$2y_1 + y_2 \geq 1.$$

Let us solve problem (D*) applying Algorithm 4.4.1.

Set $y^0 = \begin{pmatrix} 2 \\ 3 \end{pmatrix}$ (i.e., the end point of the segment).

Compute A^1, y^1, such that

$$A_1^{1T} y^0 = \min\{A_1^T y^0 \mid A_1 \in \mathbf{A}_1\},$$

which is equivalent to finding

$$\min\{2a_1 + 3a_2 \mid \begin{pmatrix} a_1 \\ a_2 \end{pmatrix} = \lambda \begin{pmatrix} 1 \\ 2 \end{pmatrix} + (1 - \lambda) \begin{pmatrix} 2 \\ 3 \end{pmatrix}, \lambda \in [0, 1]\}$$
$$= \min\{13 - 5\lambda \mid \lambda \in [0, 1]\}.$$

This minimum is attained at $\lambda = 1$; hence $A_1^1 = \begin{pmatrix} 1 \\ 2 \end{pmatrix}$

and $A^{1T} = \begin{pmatrix} 1 & 2 \\ 2 & 1 \end{pmatrix}$. Then

$$y^1 = (A^{1T})^{-1}c = \begin{pmatrix} -\frac{1}{3} & \frac{2}{3} \\ \frac{2}{3} & -\frac{1}{3} \end{pmatrix} \begin{pmatrix} 3 \\ 1 \end{pmatrix} = \begin{pmatrix} -\frac{1}{3} \\ \frac{5}{3} \end{pmatrix}.$$

Let $y^1 \neq y^0$ and then compute

$$A_1^{2T} y^0 = \min\{A_1^T y^1 \mid A_1 \in \mathbf{A}_1\},$$

which is equivalent to finding

$$\min\{-\tfrac{1}{3}a_1 + \tfrac{5}{3}a_2 \mid \begin{pmatrix} a_1 \\ a_2 \end{pmatrix} = \lambda \begin{pmatrix} 1 \\ 2 \end{pmatrix} + (1 - \lambda) \begin{pmatrix} 2 \\ 3 \end{pmatrix}, \lambda \in [0, 1]\}$$
$$= \min\{\tfrac{43}{3} - \tfrac{13}{3}\lambda \mid \lambda \in [0, 1]\}.$$

This minimum is also attained at $\lambda = 1$; hence $A_1^2 = \begin{pmatrix} 1 \\ 2 \end{pmatrix}$

and $A^{2T} = \begin{pmatrix} 1 & 2 \\ 2 & 1 \end{pmatrix}$. Then

$$y^2 = (A^{2T})^{-1}c = \begin{pmatrix} -\frac{1}{3} & \frac{2}{3} \\ \frac{2}{3} & -\frac{1}{3} \end{pmatrix} \begin{pmatrix} 3 \\ 1 \end{pmatrix} = \begin{pmatrix} -\frac{1}{3} \\ \frac{5}{3} \end{pmatrix}.$$

If $y^{2i} = y^1$ then

$$y^* = y^2 = \begin{pmatrix} -\frac{1}{3} \\ \frac{5}{3} \end{pmatrix}, \quad A^* = (A_1^2, A_2^2) = \begin{pmatrix} 1 & 2 \\ 2 & 1 \end{pmatrix}.$$

A^*, y^* is an optimal solution of dual problem (D*). Substituting A^* into (P*) we easily verify that $x^* = \begin{pmatrix} \frac{5}{3} \\ \frac{2}{3} \end{pmatrix}$ and $f(\mathbf{A}, \{b\}, \{c\}) = g(\mathbf{A}, \{b\}, \{c\}) = \frac{17}{3}$.

4.4.2 Rectangular matrix

Now, we investigate the LPSC problem (4.29) for the more general case of the rectangular matrix A of the $m \times n$-type in constraints (4.29); particularly, we consider $m < n$. Remember that there is a set $\mathbf{A} = (\mathbf{A}_1, \dots, \mathbf{A}_n)$, $\mathbf{A} \subseteq \mathbb{R}^m$; the columns A_j of the matrix A can be chosen from the sets A_j, $j = 1, 2, \dots, n$.

Suppose the basis B of columns is given. The set of indices I of the columns can be divided into basic and nonbasic indices B and N, respectively; i.e.

$$I = \{1, 2, \dots, n\} = B \cup N,$$

where B is the set of basic indices and N denotes nonbasic indices. Then the original constraints of (4.29),

$$Ax = b, \quad x \geq 0,$$

are equivalent to

$$A_B x_B + A_N x_N = b, \quad x_B \geq 0, x_N = 0;$$

i.e.,

$$A_B x_B = b, \quad x_B \geq 0, \tag{4.45}$$

where A_B is a square $m \times m$ matrix. System (4.45) has been already investigated in the previous sections.

Another problem is the question of how to find "the best basis." For this purpose, in the following algorithm we modify the well-known rules of the simplex method.

4.4.3 Finding the best basis: Algorithm

select basis B^* arbitrarily;
repeat
    $B := B^*$;
    by Algorithm 4.4.1 find A_B^* such that
    $y_B^* = (A_B^{*T})^{-1}c$;
    $x_B^* = (A_B^*)^{-1}b$;
    $q_j = (A_B^{*T}y_B^*)_j - c_j, j \in N$;
    if $q_k < 0$ for some $k \in N$ **then** introduce the kth column
    into the basis with the usual rules for selecting
    the pivot; find a new basis B^*;
    end;
until $q_k \geq 0$ for all $k \in N$
% $x^* = (x_B^*, 0)$ is the optimal solution of the problem:
% Max $\{c^T x | x \in X^= (\mathbf{A}, \{b\})\}$
% y_B^* is the optimal solution of the dual problem,
% x_B^* is the optimal B-basic solution,
% $b^T y_B^*$ is the corresponding optimal value,
% B is the best basis.

If the above algorithm stops, then the optimal solution has been reached. However, it is necessary to stress that the above-mentioned algorithms are not safe methods for solving LPSC problems in general. There are two possible obstructions.

• It may happen that $\mathbf{A}_{Bi}$ is not regular; i.e., it contains a singular matrix. Then the algorithm fails due to the nonsolvability of the subproblem, although it need not necessarily mean that the original problem does not have an optimal solution.

• If inequality $x_B^* \geq 0$ is violated for a current basis, we must carry out necessary steps in order to satisfy nonnegativity of the basic solution. In [98], it is proposed that the negative components of the basic solution in x_B^* are fixed to the value zero. Of course, such a modification of the algorithm acquires a heuristic character.

It is an open problem how to resolve the above obstructions. This problem exceeds, however, the scope of this chapter.

4.5 Conclusion

In this chapter, new problems of linear programming with the coefficients not given as strict but variable within the given sets have been investigated. Such

problems—called linear programming problems with set coefficients (LPSC)—
are sometimes referred to as "post-optimal analysis"; sometimes they are included in the parametric programming framework. The nature of these problems is, however, varied. First, we have introduced the linear optimization
problem with set coefficients, then weak, strict and strong feasibility concepts
have been defined and discussed. The major part of the chapter has been
devoted to the problem of duality of LPSC problems. Under general assumptions, the usual form of the weak duality theorem has been derived. Based on
the results of [98], the strong duality theorem has also been formulated and
proven. The last part of the work has dealt with methods for solving LPSC
problems. Two such algorithms have been proposed and their efficiency has
been discussed. The proposed algorithms are in fact a generalization of the
well-known simplex method. An illustrative example has been presented and
discussed.

5

Fuzzy linear optimization

J. Ramík

5.1 Introduction

In mathematical optimization problems preferences between alternatives are described by means of objective functions on a given set of alternatives. The values of the objective function describe effects from the alternatives; the more preferable alternatives have higher values than the less preferable ones. For example, in economic problems these values may reflect profits obtained in various means of production. The set of feasible alternatives in mathematical optimization problems is described by means of constraints—equations or inequalities—representing relevant relationships between alternatives. The results of the analysis depend largely upon how adequately various factors of the real system are reflected in the description of the objective function(s) and the constraints.

Mathematical formulation of the objective function and of the constraints in mathematical optimization problems usually includes some parameters; e.g., in problems of resource allocation the parameters may represent economic values such as costs of various types of production, shipment costs, etc.

The values of such parameters depend on multiple factors usually not included in the formulation of the problem. Trying to make the model more representative, we often include the corresponding complex relations, causing the model to become more cumbersome and analytically unsolvable. Some attempts to increase "precision" of the model will be of no practical value due to the impossibility of measuring the parameters accurately. On the other hand, the model with fixed values of its parameters may be too crude, since these values are often chosen in an arbitrary way.

An alternative approach is based on introducing into the model a more adequate representation of expert understanding of the nature of the parameters in an adequate form. In Chapters 2 and 3 it has been done in the form of intervals; in Chapter 4 convex sets or convex polyhedral sets have been considered. Here, the parameters can be expressed in a more general form of fuzzy subsets of their possible values. In this way we obtain a new type of mathematical optimization problem containing fuzzy parameters. Considering linear optimization problems such treatment forms the essence of *fuzzy linear programming* (FLP) investigated in this chapter. As we show, for a special form of fuzzy parameters, crisp parameters, i.e. the usual real numbers, the formulation of the linear optimization problem coincides with the corresponding formulations from the preceding chapters.

FLP problems and related ones have been extensively analyzed in many works published in papers and books displaying a variety of formulations and approaches. Most approaches to FLP problems are based on the straightforward use of the intersection of fuzzy sets representing goals and constraints. The resulting membership function is then maximized. This approach has been mentioned by Bellman and Zadeh in [13]. Later on many papers were devoted to the problem of linear programming with fuzzy parameters, known under different names, mostly as fuzzy linear programming, but sometimes as possibilistic LP, flexible linear programming, vague linear programming, inexact linear programming, etc. For an extensive bibliography, see the overview in paper [58].

Here we present an approach based on a systematic extension of the traditional formulation of the LP problem. This approach is based on previous works of the authors of this book; see [118], [119], [120], [121], [124], [125], [126], [128], [129], [130], [131], [132], [136], [137], and also on the works of many other authors, e.g., [22], [26], [27], [33], [57], [58], [59], [76], [77], [78], [81], [82], [113], [114], [133], [169], [170], [174], [195], [197].

In this chapter, we demonstrate that FLP essentially differs from stochastic programming; FLP has its own structure and tools for investigating broad classes of optimization problems. FLP is also different from parametric LP. Problems of parametric LP are in essence deterministic optimization problems with special variables called parameters. The main interest in parametric LP is focused on finding functional relationships between the values of parameters and optimal solutions of the LP problem.

An appropriate treatment of FLP problems requires proper application of special tools in a logically consistent manner. An important role in this treatment is played by generalized concave membership functions and fuzzy relations. The following treatment is based on the substance partly investigated in [135].

First we formulate an optimization problem and, particularly, the FLP problem associated with a collection of instances of the classical LP problem. After that we define a feasible solution of the FLP problem and deal with the problem of "optimal solution" of FLP problems. Two approaches are introduced: the first one, satisficing solution, is based on external goals modeled by fuzzy quantities; the second approach is based on the concept of an efficient (nondominated) solution. Then our interest is focused on the problem of duality in FLP problems. Finally, we also investigate the multicriteria case. We formulate the fuzzy multicriteria linear programming problem, define a compromise solution and derive appropriate results. The chapter closes with a numerical example.

5.2 Fuzzy sets, fuzzy quantities

In this section we summarize basic notions and results from fuzzy set theory that are useful in this chapter. Throughout this section, X is a nonempty set.

Definition 5.1. *A fuzzy subset A of X is the family of subsets $A_\alpha \subseteq X$, where $\alpha \in [0, 1]$, satisfying the following properties.*

$$A_0 = X, \tag{5.1}$$

$$A_\beta \subseteq A_\alpha \quad \text{whenever } 0 \leq \alpha < \beta \leq 1, \tag{5.2}$$

$$A_\beta = \bigcap_{0 \leq \alpha < \beta} A_\alpha. \tag{5.3}$$

A fuzzy subset A of X is called a fuzzy set. *The class of all fuzzy subsets of X is denoted by $\mathcal{F}(X)$.*

Definition 5.2. *Let A be a subset of X. The fuzzy subset $\{A_\alpha\}_{\alpha \in [0,1]}$ of X defined by $A_\alpha = A$ for all $\alpha \in (0, 1]$ is called a* crisp fuzzy subset *of X generated by A. A fuzzy subset of X generated by some $A \subseteq X$ is called a* crisp fuzzy subset *of X or briefly a* crisp subset *of X.*

In Definition 5.2 crisp fuzzy subsets of X and "classic" subsets of X are in one-to-one correspondence. In this way, "classic" subsets of X are isomorphically embedded into fuzzy subsets of X.

Definition 5.3. *Let $A = \{A_\alpha\}_{\alpha \in [0,1]}$ be a fuzzy subset of X. The $\mu_A : X \to [0, 1]$ defined by*

$$\mu_A(x) = \sup\{\alpha \mid \alpha \in [0, 1], x \in A_\alpha\} \tag{5.4}$$

is called the membership function *of A, and the value $\mu_A(x)$ is called the* membership degree *of x in the fuzzy set A.*

Notice that the membership function of a crisp fuzzy subset of X is equal to the characteristic function of the corresponding set.

Definition 5.4. *Let A be a fuzzy subset of X. The* core *of A, Core(A), is defined by*

$$\text{Core}(A) = \{x \in X \mid \mu_A(x) = 1\}.$$

The complement *of A, $\mathcal{C}A$, is defined by*

$$\mu_{\mathcal{C}A}(x) = 1 - \mu_A(x). \tag{5.5}$$

If the core of A is nonempty, then A is said to be normalized. *The* support *of A, Supp(A), is defined by*

$$Supp(A) = Cl(\{x \in X \mid \mu_A(x) > 0\}).$$

Here, by Cl *we denote the topological closure.*
The height *of A,* Hgt(A), *is defined by*

$$Hgt(A) = \sup\{\mu_A(x) \mid x \in X\}.$$

The upper-level set *of the membership function μ_A of A at $\alpha \in [0,1]$ is denoted by $[A]_\alpha$ and called the α-cut of A; that is,*

$$[A]_\alpha = \{x \in X \mid \mu_A(x) \geq \alpha\}. \tag{5.6}$$

The strict upper-level set *of the membership function μ_A of A at $\alpha \in [0,1)$ is denoted by $(A)_\alpha$ and called the* strict α-cut *of A; that is,*

$$(A)_\alpha = \{x \in X \mid \mu_A(x) > \alpha\}. \tag{5.7}$$

Note that if A is normalized, then $Hgt(A) = 1$, but not vice versa.

Definition 5.5. *Let $X \subseteq \mathbb{R}^m$, the m-dimensional Euclidean space. A fuzzy subset $A = \{A_\alpha\}_{\alpha \in [0,1]}$ of X is called* closed, bounded, compact *or* convex *if A_α is a closed, bounded, compact or convex subset of X for every $\alpha \in (0,1]$, respectively.*

In the following two propositions, we show that the family generated by the upper level sets of a function $\mu : X \to [0,1]$, satisfies conditions (5.1)–(5.3); thus, it generates a fuzzy subset of X and the membership function μ_A defined by (5.4) coincides with μ. Moreover, for a given fuzzy set $A = \{A_\alpha\}_{\alpha \in [0,1]}$, every α-cut $[A]_\alpha$ given by (5.6) coincides with the corresponding A_α. The proofs are easy and can be found in [135].

Proposition 5.6. *Let $\mu : X \to [0,1]$ be a function and let $A = \{A_\alpha\}_{\alpha \in [0,1]}$ be a family of its upper-level sets. Then A is a fuzzy subset of X and μ is the membership function of A.*

Proposition 5.7. *Let $A = \{A_\alpha\}_{\alpha \in [0,1]}$ be a fuzzy subset of X and let $\mu_A : X \to [0,1]$ be the membership function of A. Then for each $\alpha \in [0,1]$ the α-cut $[A]_\alpha$ is equal to A_α.*

Now, we investigate fuzzy subsets of the real line; we set $X = \mathbb{R}$ and $\mathcal{F}(X) = \mathcal{F}(\mathbb{R})$.

Definition 5.8.

(i) *A fuzzy set $A = \{A_\alpha\}_{\alpha \in [0,1]}$ is called a* fuzzy interval *if for all $\alpha \in [0,1]$: A_α is nonempty and a convex subset of $\mathbb{R}$. The set of all fuzzy intervals is denoted by $\mathcal{F}_I(\mathbb{R})$.*
(ii) *A fuzzy interval A is called a* fuzzy number *if its core is a singleton. The set of all fuzzy numbers is denoted by $\mathcal{F}_N(\mathbb{R})$.*

Notice that the membership function $\mu_A : \mathbb{R} \to [0,1]$ of a fuzzy interval A is quasiconcave on $\mathbb{R}$. The following definitions are useful.

Definition 5.9. *Let $X \subseteq \mathbb{R}$. A function $f : \mathbb{R} \to [0,1]$ is called*

(i)   Quasiconcave on X *if*

$$f(\lambda x + (1 - \lambda)y) \geq \min\{f(x), f(y)\},$$

for every $x, y \in X$ and every $\lambda \in (0,1)$ with $\lambda x + (1 - \lambda)y \in X$;
(ii)   Strictly quasiconcave on X *if*

$$f(\lambda x + (1 - \lambda)y) > \min\{f(x), f(y)\}, \tag{5.8}$$

for every $x, y \in X$, $x \neq y$ and every $\lambda \in (0,1)$ with $\lambda x + (1 - \lambda)y \in X$;
(iii)   Semistrictly quasiconcave *on X if f is quasiconcave on X and (5.8) holds for every $x, y \in X$ and every $\lambda \in (0,1)$ with $\lambda x + (1 - \lambda)y \in X$, $f(\lambda x + (1 - \lambda)y) > 0$ and $f(x) \neq f(y)$.*

Notice that membership functions of crisp subsets of $\mathbb{R}$ are quasiconcave, but not stricly quasiconcave; they are, however, semistrictly quasiconcave on $\mathbb{R}$.

Definition 5.10. *A fuzzy subset A of $\mathbb{R}$ is called the* fuzzy quantity *if A is normal and compact with semistrictly quasiconcave membership function μ_A. The set of all fuzzy quantities is denoted by $\mathcal{F}_0(\mathbb{R})$.*

By the definition $\mathcal{F}_0(\mathbb{R}) \subseteq \mathcal{F}_I(\mathbb{R})$, moreover, $\mathcal{F}_0(\mathbb{R})$ contains crisp (real) numbers, crisp intervals, triangular fuzzy numbers, bell-shaped fuzzy numbers etc.

Example 5.11. (Gaussian fuzzy number) Let $a \in \mathbb{R}$, $\gamma \in (0, +\infty)$, and let for all $x \in \mathbb{R}$

$$G(x) = e^{-\frac{x^2}{\gamma}}.$$

Then the function μ_A given by

$$\mu_A(x) = G(x - a) = e^{-\frac{(x-a)^2}{\gamma}},$$

is the membership function of a fuzzy set A where A is a fuzzy number according to Definition 5.8. Notice that the Gaussian fuzzy number is compact; hence it is a fuzzy quantity.

5.3 Fuzzy relations

Now, let X and Y be nonempty sets. In set theory, a *binary relation* R between the elements of the sets X and Y is defined as a subset of the Cartesian product $X \times Y$; that is, $R \subseteq X \times Y$.

A *valued relation* R on $X \times Y$ is a fuzzy subset of $X \times Y$. A *valued relation* R *on* X is a valued relation on $X \times X$.

Any binary relation R, $R \subseteq X \times Y$, is isomorphically embedded into the class of valued relations by its characteristic function χ_R, which is its membership function. In this sense, any binary relation is valued.

Let R be a valued relation on $X \times Y$. In FLP problems, we consider fuzzy relations assigning to every pair of fuzzy subsets a real number from interval $[0, 1]$. In other words, we consider valued relations $\tilde{R}$ on $\mathcal{F}(X) \times \mathcal{F}(Y)$ such that $\mu_{\tilde{R}} : \mathcal{F}(X) \times \mathcal{F}(Y) \to [0, 1]$.

Convention: The elements $x \in X$ and $y \in Y$ are considered as fuzzy subsets of X and Y with the characteristic functions χ_x and χ_y as the membership functions. In this way we obtain the isomorphic embedding of X into $\mathcal{F}(X)$ and Y into $\mathcal{F}(Y)$, and in this sense we write $X \subseteq \mathcal{F}(X)$ and $Y \subseteq \mathcal{F}(Y)$, respectively.

Evidently, the usual binary relations $=, <$ and $\leq$ can be understood as the valued relations.

Now, we define fuzzy relations that are used for comparing the left and right sides of the constraints in optimization problems.

Definition 5.12. *A fuzzy subset of $\mathcal{F}(X) \times \mathcal{F}(Y)$ is called a fuzzy relation on $X \times Y$. The set of all fuzzy relations on $\mathcal{F}(X) \times \mathcal{F}(Y)$ is denoted by $\mathcal{F}(\mathcal{F}(X) \times \mathcal{F}(Y))$. A fuzzy relation on $X \times X$ is called a fuzzy relation on X.*

Definition 5.13. *Let R be a valued relation on $X \times Y$. A fuzzy relation $\tilde{R}$ on $X \times Y$ given by the membership function $\mu_{\tilde{R}} : \mathcal{F}(X) \times \mathcal{F}(Y) \to [0, 1]$ is called a fuzzy extension of relation R, if for each $x \in X$, $y \in Y$, it holds*

$$\mu_{\tilde{R}}(x, y) = \mu_R(x, y). \tag{5.9}$$

On the left side of (5.9), x and y are understood as fuzzy subsets of X and Y defined by the membership functions identical with the characteristic functions of singletons $\{x\}$ and $\{y\}$, respectively.

Definition 5.14. *Let $\Psi : \mathcal{F}(X \times Y) \to \mathcal{F}(\mathcal{F}(X) \times \mathcal{F}(Y))$ be a mapping. Let for all $R \in \mathcal{F}(X \times Y)$, $\Psi(R)$ be a fuzzy extension of relation R. Then the mapping Ψ is called a fuzzy extension of valued relations.*

Definition 5.15. *Let $\Phi, \Psi : \mathcal{F}(X \times Y) \to \mathcal{F}(\mathcal{F}(X) \times \mathcal{F}(Y))$ be mappings. We say that the mapping Φ is dual to Ψ, if*

$$\Phi(\mathcal{C}R) = \mathcal{C}\Psi(R) \tag{5.10}$$

holds for all $R \in \mathcal{F}(X \times Y)$. For Φ dual to Ψ, $R \in \mathcal{F}(X \times Y)$ a valued relation, the fuzzy relation $\Phi(R)$ is called dual to fuzzy relation $\Psi(R)$.

Proposition 5.16. *A mapping Φ is dual to Ψ, if and only if the mapping Ψ is dual to Φ.*

Proof. The proposition follows easily from (5.10), (5.5) and from the identity

$$CCR = R.$$

□

The analogous statement holds for the dual fuzzy relations $\Phi(R)$ and $\Psi(R)$.

Now, we define special mappings, important fuzzy extensions of valued relations. The concept of t-norm, and t-conorm are useful.

A class of functions $T : [0,1]^2 \to [0,1]$ that are commutative, associative, nondecreasing in every variable and satisfy the following boundary condition

$$T(a,1) = a \quad \text{for all } a \in [0,1],$$

is called the *triangular norms* or *t-norms*. The four most popular examples of t-norms are defined as follows.

$$T_M(a,b) = \min\{a,b\},$$
$$T_P(a,b) = a.b,$$
$$T_L(a,b) = \max\{0, a+b-1\}.$$
$$T_D(a,b) = \min\{a,b\} \text{ if } \max\{a,b\} = 1$$
$$\qquad\quad = 0 \qquad\qquad \text{otherwise.}$$

They are called minimum t-norm T_M, product t-norm T_P, Lukasiewicz t-norm T_L and drastic product T_D.

A class of functions closely related to the class of t-norms is the class of functions $S : [0,1]^2 \to [0,1]$ that are commutative, associative, nondecreasing in every variable and satisfy the following boundary condition,

$$S(a,0) = a \quad \text{for all } a \in [0,1].$$

The functions that satisfy all these properties are called the *triangular conorms* or *t-conorms*; see, e.g., [73]. For example, the functions S_M, S_P, S_L defined for $a, b \in [0,1]$ by

$$S_M(a,b) = \max\{a,b\},$$
$$S_P(a,b) = a+b-a\cdot b,$$
$$S_L(a,b) = \min\{1, a+b\},$$
$$S_D(a,b) = \begin{cases} \max\{a,b\} & \text{if } \min\{a,b\} = 0, \\ 1 & \text{otherwise} \end{cases}$$

are the t-conorms. S_M, S_P, S_L and S_D are often called the maximum, probabilistic sum, bounded sum and drastic sum, respectively. It can easily be verified that for each t-norm T, the function $T^* : [0,1]^2 \to [0,1]$ defined for all $a, b \in [0,1]$ by

$$T^*(a,b) = 1 - T(1-a, 1-b) \tag{5.11}$$

is a t-conorm. The converse statement is also true. Namely, if S is a t-conorm, then the function $S^* : [0,1]^2 \to [0,1]$ defined for all $a, b \in [0,1]$ by

$$S^*(a,b) = 1 - S(1-a, 1-b) \tag{5.12}$$

is a t-norm. The t-conorm T^* and t-norm S^*, are called *dual* to the t-norm T and t-conorm S, respectively. It may easily be verified that

$$T_M^* = S_M, \quad T_P^* = S_P, \quad T_L^* = S_L, \quad T_D^* = S_D.$$

A triangular norm T is said to be *strict* if it is continuous and strictly monotone. It is said to be *Archimedean* if for all $x, y \in (0,1)$ there exists a positive integer n such that $T^{n-1}(x, \ldots, x) < y$. Here, by commutativity and associativity we can define the extension to more than two arguments by the formula

$$T^{n-1}(x_1, x_2, \ldots, x_n) = T(T^{n-2}(x_1, x_2, \ldots, x_{n-1}), x_n), \tag{5.13}$$

where $T^1(x_1, x_2) = T(x_1, x_2)$.

Notice that if T is strict, then T is Archimedean.

Definition 5.17. *An additive generator of a t-norm T is a strictly decreasing function $f : [0,1] \to [0, +\infty]$ which is right continuous at 0, satisfies $f(1) = 0$, and is such that for all $x, y \in [0,1]$ we have*

$$f(x) + f(y) \in \mathrm{Ran}(f) \cup [f(0), +\infty], \tag{5.14}$$

$$T(x,y) = f^{(-1)}(f(x) + f(y)), \tag{5.15}$$

where $\mathrm{Ran}(f) = \{ y \in \mathbb{R} \mid y = f(x), x \in [0,1] \}$.

Triangular norms (t-conorms) constructed by means of additive (multiplicative) generators are always Archimedean. This property and some other properties of t-norms are summarized in [73].

Definition 5.18. *Let T be a t-norm and S be a t-conorm. Let R be a valued relation on X. Fuzzy extensions $\Phi^T(R)$ and $\Phi^S(R)$ of a valued relation R on X defined for all fuzzy sets A, B with the membership functions $\mu_A : X \to [0,1]$, $\mu_B : Y \to [0,1]$, respectively, by*

$$\mu_{\Phi^T(R)}(A,B) = \sup\{T(\mu_R(x,y), T(\mu_A(x), \mu_B(y))) | x, y \in X\}, \tag{5.16}$$

$$\mu_{\Phi^S(R)}(A,B) = \inf\{S(S(1 - \mu_A(x), 1 - \mu_B(y)), \mu_R(x,y)) | x, y \in X\}, \tag{5.17}$$

are called a T-fuzzy extension of relation R and S-fuzzy extension of relation R, respectively.

It can be easily verified that the T-fuzzy extension of relation R and S-fuzzy extension of relation R are fuzzy extensions of relation R given by Definition 5.13.

In the following proposition we prove a duality result between fuzzy extensions of valued relations. In a special case, particularly $T = \min$ and $S = \max$, the analogous results can be also found in [56].

Proposition 5.19. *Let T be a t-norm and S be a t-conorm dual to T. Then Φ^T is dual to Φ^S.*

Proof. Let $R \in \mathcal{F}(X \times Y)$; we have to prove (5.10); i.e.,

$$\Phi^T(\mathcal{C}R) = \mathcal{C}\Phi^S(R). \tag{5.18}$$

To prove (5.18), let $A \in \mathcal{F}(X)$ and $B \in \mathcal{F}(Y)$. We have to show that

$$\mu_{\Phi^T(\mathcal{C}R)}(A, B) = \mu_{\mathcal{C}\Phi^S(R)}(A, B).$$

By definition (5.16) and by duality of T and S (see (5.11), (5.12)), we obtain

$$
\begin{aligned}
\mu_{\Phi^T(\mathcal{C}R)}&(A, B) \\
&= \sup\{T(T(\mu_A(x), \mu_B(y)), \mu_{\mathcal{C}R}(x, y)) \mid x \in X, \, y \in Y\} \\
&= \sup\{1 - S(1 - T(\mu_A(x), \mu_B(y)), \mu_R(x, y)) \mid x \in X, \, y \in Y\} \\
&= 1 - \inf\{S(S(\mu_{\mathcal{C}A}(x), \mu_{\mathcal{C}B}(y)), \mu_R(x, y)) \mid x \in X, \, y \in Y\} \\
&= 1 - \mu_{\Phi^S(R)}(A, B) = \mu_{\mathcal{C}\Phi^S(R)}(A, B).
\end{aligned}
$$

This is the required result.   □

Definition 5.20. *Let R be $\leq$, i.e., R be a classical binary relation "less or equal" on $\mathbb{R}$; let $T = \min$ and $S = \max$. We denote $\Phi^T(R)$ and $\Phi^S(R)$ from (5.16) and (5.17) by $\widetilde{\leq}^{\min}$ and $\widetilde{\leq}^{\max}$, respectively. From (5.16) and (5.17) we obtain two fuzzy extensions of relation $\leq$ by*

$$\mu_{\widetilde{\leq}^{\min}}(A, B) = \sup\{\min(\mu_A(x), \mu_B(y), \mu_R(x, y)) | x, y \in \mathbb{R}\}, \tag{5.19}$$

$$\mu_{\widetilde{\leq}^{\max}}(A, B) = \inf\{\max(1 - \mu_A(x), 1 - \mu_B(y), \mu_R(x, y)) | x, y \in \mathbb{R}\}. \tag{5.20}$$

We equivalently write $A\widetilde{\leq}^{\min}B$ and $A\widetilde{\leq}^{\max}B$, instead of $\mu_{\widetilde{\leq}^{\min}}(A, B)$ and $\mu_{\widetilde{\leq}^{\max}}(A, B)$, respectively. By $A\widetilde{\geq}^{\min}B$ we mean $B\widetilde{\leq}^{\min}A$.

The following results are crucial for studying FLP problems.

Theorem 5.21. *Let R be $\leq$, let $T = \min$ and $S = \max$. Let $A, B \in \mathcal{F}(\mathbb{R})$ be normal and compact fuzzy sets, $\alpha \in (0, 1)$. Then*
(i) $\mu_{\widetilde{\leq}^{\min}}(A, B) \geq \alpha$ if and only if $\inf[A]_\alpha \leq \sup[B]_\alpha$,
(ii) $\mu_{\widetilde{\leq}^{\max}}(A, B) \geq \alpha$ if and only if $\sup(A)_{1-\alpha} \leq \inf(B)_{1-\alpha}$.

Proof. First we prove (i). Let $\alpha \in (0,1)$ and $\mu_{\lesssim \min}(A, B) \geq \alpha$. Then by (5.19) we get

$$\sup\{\min(\mu_A(x), \mu_B(y))|x \leq y\} \geq \alpha.$$

Since $[A]_\alpha$ and $[B]_\alpha$ are nonempty and compact, we obtain $\inf[A]_\alpha \leq \sup[B]_\alpha$.

On the other hand, let $\alpha \in (0,1)$ and $\inf[A]_\alpha \leq \sup[B]_\alpha$. By compactness there exist $x' \in [A]_\alpha$ and $y' \in [B]_\alpha$ such that $x' \leq y'$. Then $\mu_A(x') \geq \alpha, \mu_B(y) \geq \alpha$, $\min(\mu_A(x') and \mu_B(y')) \geq \alpha$; consequently

$$\sup\{\min(\mu_A(x), \mu_B(y))|x \leq y\} \geq \alpha, \quad i.e. \ \mu_{\lesssim \min}(A, B) \geq \alpha.$$

Secondly, we prove (ii).

Let $\alpha \in (0,1)$ and $\mu_{\lesssim \max}(A, B) \geq \alpha$. Then by (5.20) we get

$$\inf\{\max(1 - \mu_A(x), 1 - \mu_B(y), \mu_R(x, y))|x, y \in \mathbb{R}\} \geq \alpha.$$

Clearly, this inequality is equivalent to

$$\sup\{\min(\mu_A(x), \mu_B(y))|x > y\} \leq 1 - \alpha. \tag{5.21}$$

Take arbitrary $x' \in (A)_{1-\alpha}$ and $y' \in (B)_{1-\alpha}$. Assume that $x' > y'$; then $\min(\mu_A(x'), \mu_B(y')) > 1 - \alpha$, which contradicts (5.21). Thus $x' \leq y'$ holds for any couple $x' \in (A)_{1-\alpha}$, $y' \in (B)_{1-\alpha}$ that gives $\sup(A)_{1-\alpha} \leq \inf(B)_{1-\alpha}$.

On the other hand, let $\alpha \in (0,1)$ and $\sup(A)_{1-\alpha} \leq \inf(B)_{1-\alpha}$. Take arbitrary $x', y' \in \mathbb{R}$ such that $x' > y'$. Then either $x' \notin (A)_{1-\alpha}$ or $y' \notin (B)_{1-\alpha}$; otherwise $x' \leq y'$. Then $\min(\mu_A(x'), \mu_B(y')) \leq 1 - \alpha$; consequently $\sup\{\min(\mu_A(x), \mu_B(y))|x \leq y\} \leq 1 - \alpha$, which is equivalent to $\mu_{\lesssim \max}(A, B) \geq \alpha$. $\qquad\square$

Let T be a t-norm and S be a t-conorm.

Definition 5.22.

(1) A mapping $\Psi^{T,S} : \mathcal{F}(X \times Y) \to \mathcal{F}(\mathcal{F}(X) \times \mathcal{F}(Y))$ is defined for every valued relation $R \in \mathcal{F}(X \times Y)$ and for all fuzzy sets $A \in \mathcal{F}(X)$, $B \in \mathcal{F}(Y)$ by

$$
\mu_{\Psi^{T,S}(R)}(A, B)
$$
$$
= \sup\{\inf\{T(\mu_A(x), S(\mu_{CB}(y), \mu_R(x, y))) \mid y \in Y\} \mid x \in X\}. \tag{5.22}
$$

(2) A mapping $\Psi_{T,S} : \mathcal{F}(X \times Y) \to \mathcal{F}(\mathcal{F}(X) \times \mathcal{F}(Y))$ is defined for every valued relation $R \in \mathcal{F}(X \times Y)$ and for all fuzzy sets $A \in \mathcal{F}(X)$, $B \in \mathcal{F}(Y)$ by

$$
\mu_{\Psi_{T,S}(R)}(A, B)
$$
$$
= \inf\{\sup\{S(T(\mu_A(x), \mu_R(x, y)), \mu_{CB}(y)) \mid x \in X\} \mid y \in Y\}. \tag{5.23}
$$

(3) A mapping $\Psi^{S,T} : \mathcal{F}(X \times Y) \to \mathcal{F}(\mathcal{F}(X) \times \mathcal{F}(Y))$ is defined for every valued relation $R \in \mathcal{F}(X \times Y)$ and for all fuzzy sets $A \in \mathcal{F}(X)$, $B \in \mathcal{F}(Y)$ by

$$\mu_{\Psi^{S,T}(R)}(A, B) \\ = \sup\{\inf\{T(S(\mu_{CA}(x), \mu_R(x, y)), \mu_B(y)) \mid x \in X\} \mid y \in Y\}.$$
(5.24)

(4) A mapping $\Psi_{S,T} : \mathcal{F}(X \times Y) \to \mathcal{F}(\mathcal{F}(X) \times \mathcal{F}(Y))$ is defined for every valued relation $R \in \mathcal{F}(X \times Y)$ and for all fuzzy sets $A \in \mathcal{F}(X)$, $B \in \mathcal{F}(Y)$ by

$$\mu_{\Psi_{S,T}(R)}(A, B) \\ = \inf\{\sup\{S(\mu_{CA}(x), T(\mu_B(y), \mu_R(x, y))) \mid y \in Y\} \mid x \in X\}.$$
(5.25)

The previous four fuzzy relations are also fuzzy extensions of valued relations by Definition 5.14.

5.4 Fuzzy linear optimization problems

Now, we turn to optimization theory and consider the following *optimization problem*,

$$\text{maximize (minimize) } f(x)$$
$$\text{subject to} \qquad x \in X,$$
(5.26)

where f is a real-valued function on $\mathbb{R}^n$ called the *objective function* and X is a nonempty subset of $\mathbb{R}^n$ given by means of real-valued functions $g_1, g_2, \ldots, g_m$ on $\mathbb{R}^n$, the set of all solutions of the system

$$g_i(x) = b_i, \quad i = 1, 2, \ldots, m_1,$$
$$g_i(x) \leq b_i, \quad i = m_1 + 1, m_1 + 2, \ldots, m,$$
$$x_j \geq 0, \ j = 1, 2, \ldots, n,$$

called the *constraints*. The elements of X are called *feasible solutions* of (5.26), and the feasible solution x^* where f attains its global maximum (or minimum) over X is called the *optimal solution*.

Most frequent optimization problems are linear ones. In this chapter we are concerned with the fuzzy linear programming problem related to linear programming problems in the following form.

Let $\mathcal{M} = \{1, 2, \ldots, m\}$ and $\mathcal{N} = \{1, 2, \ldots, n\}$ where m and n are positive integers. For each $c = (c_1, c_2, \ldots, c_n)^T \in \mathbb{R}^n$ and $a_i = (a_{i1}, a_{i2}, \ldots, a_{in})^T \in \mathbb{R}^n$, $i \in \mathcal{M}$, the functions $f(\cdot, c)$ and $g(\cdot, a_i)$ defined on $\mathbb{R}^n$ by

$$f(x, c_1, \ldots, c_n) = c_1 x_1 + \cdots + c_n x_n,$$
(5.27)

$$g_i(x, a_{i1}, \ldots, a_{in}) = a_{i1} x_1 + \cdots + a_{in} x_n, \quad i \in \mathcal{M},$$
(5.28)

are linear on $\mathbb{R}^n$. For each $c \in \mathbb{R}^n$ and $a_i \in \mathbb{R}^n$, $i \in \mathcal{M}$, we consider the *linear programming problem* (classical LP),

$$\text{maximize (minimize) } c_1 x_1 + \cdots + c_n x_n$$
$$\text{subject to} \qquad a_{i1} x_1 + \cdots + a_{in} x_n \leq b_i, \quad i \in \mathcal{M}, \qquad (5.29)$$
$$x_j \geq 0, \quad j \in \mathcal{N}.$$

The set of all feasible solutions of problem (5.29) is denoted by X; that is,

$$X = \{x \in \mathbb{R}^n \mid a_{i1} x_1 + \cdots + a_{in} x_n \leq b_i, i \in \mathcal{M}, x_j \geq 0, j \in \mathcal{N}\}. \qquad (5.30)$$

Assumptions and remarks.

1. Let f, g_i be linear functions defined by (5.27), (5.28), respectively. From now on, the parameters c_j, a_{ij} and b_i are considered as *fuzzy quantities*, that is, normal and compact fuzzy subsets of the Euclidean space $\mathbb{R}$ with semi-strictly quasiconcave membership function; see Definition 5.10. This assumption makes it possible to include classical LP problems in fuzzy LP ones. The fuzzy quantities are denoted with the tilde above the corresponding symbol. We also have $\mu_{\tilde{c}_j} : \mathbb{R} \to [0,1]$, $\mu_{\tilde{a}_{ij}} : \mathbb{R} \to [0,1]$ and $\mu_{\tilde{b}_i} : \mathbb{R} \to [0,1]$, $i \in \mathcal{M}$, $j \in \mathcal{N}$, membership functions of the fuzzy parameters $\tilde{c}_j$, $\tilde{a}_{ij}$ and $\tilde{b}_i$, respectively. The crisp quantities are not denoted with the tilde.

2. Let $\tilde{R}_i$, $i \in \mathcal{M}$, be fuzzy relations on $\mathbb{R}$. They are used for "comparing the left and right sides" of the constraints. Primarily, we study the case of $\tilde{R}_i = \tilde{R}$, for all $i \in \mathcal{M}$; i.e., all fuzzy relations in the constraints are the same.

3. The "optimization" i.e., "maximization" or "minimization" of the objective function requires special treatment, as the set of fuzzy values of the objective function is not linearly ordered. In order to "maximize" the objective function we define a suitable concept of "optimal solution". It is done by two distinct approaches: applying the first approach, an exogenously given fuzzy goal $\tilde{d} \in \mathcal{F}(\mathbb{R})$ and special fuzzy relation $\tilde{R}_0$ on $\mathbb{R}$ is introduced. In the second approach we define an α-efficient (α-nondominated) solution of the FLP problem. Some other approaches can be found in the literature; see [39], [33], [136].

The *fuzzy linear programming problem (FLP problem)* associated with the LP problem (5.29) is defined as follows,

$$\text{"maximize" ("minimize") } \tilde{c}_1 x_1 \tilde{+} \cdots \tilde{+} \tilde{c}_n x_n$$
$$\text{subject to} \qquad (\tilde{a}_{i1} x_1 \tilde{+} \cdots \tilde{+} \tilde{a}_{in} x_n) \tilde{R}_i \tilde{b}_i, \quad i \in \mathcal{M}, \qquad (5.31)$$
$$x_j \geq 0, \quad j \in \mathcal{N}.$$

Here, $\tilde{R}_i$, $i \in \mathcal{M}$, are fuzzy relations on $\mathbb{R}$. The objective function values and the left-hand side values of the constraints of (5.31) are obtained by the *extension principle*:

For given $\tilde{c}_1, \ldots, \tilde{c}_n \in \mathcal{F}_0(\mathbb{R})$, $\tilde{f}(x, \tilde{c}_1, \ldots, \tilde{c}_n)$ is the fuzzy extension of $f(x, c_1, \ldots, c_n)$ with the membership function defined for each $t \in \mathbb{R}$ by

$$
\mu_{\tilde{f}}(t) = \begin{cases} \sup \left\{ T(\mu_{\tilde{c}_1}(c_1), \ldots, \mu_{\tilde{c}_n}(c_n)) \ \middle| \ \begin{array}{l} c_1, \ldots, c_n \in \mathbb{R}, \\ c_1 x_1 + \cdots + c_n x_n = t \end{array} \right\} \\ \qquad\qquad\qquad\qquad\qquad\qquad \text{if } f^{-1}(x; t) \neq \emptyset, \\ 0 \qquad\qquad\qquad\qquad\qquad\qquad \text{otherwise}, \end{cases}
$$

(5.32)

where $f^{-1}(x, t) = \{(c_1, \ldots, c_n)^T \in \mathbb{R}^n | f(x, c_1, \ldots, c_n) = t\}$.

Particularly, for $f(x, c_1, \ldots, c_n) = c_1 x_1 + \cdots + c_n x_n$, the fuzzy set $\tilde{f}(x, \tilde{c}_1, \ldots, \tilde{c}_n)$ is denoted as $\tilde{c}_1 x_1 \tilde{+} \cdots \tilde{+} \tilde{c}_n x_n$; i.e.,

$$
\tilde{f}(x, \tilde{c}_1, \ldots, \tilde{c}_n) = \tilde{c}_1 x_1 \tilde{+} \cdots \tilde{+} \tilde{c}_n x_n. \tag{5.33}
$$

Similarly, the membership function of $\tilde{g}_i(x, \tilde{a}_{i1}, \ldots, \tilde{a}_{i1})$ is defined for each $t \in \mathbb{R}$ by

$$
\mu_{\tilde{g}_i}(t) = \begin{cases} \sup \left\{ T(\mu_{\tilde{a}_{i1}}(a_1), \ldots, \mu_{\tilde{a}_{in}}(a_n)) \ \middle| \ \begin{array}{l} a_1, \ldots, a_n \in \mathbb{R}, \\ a_1 x_1 + \cdots + a_n x_n = t \end{array} \right\} \\ \qquad\qquad\qquad\qquad\qquad\qquad \text{if } g_i^{-1}(x; t) \neq \emptyset, \\ 0 \qquad\qquad\qquad\qquad\qquad\qquad \text{otherwise}, \end{cases}
$$

(5.34)

where

$$
g_i^{-1}(x, t) = \{(a_1, \ldots, a_n)^T \in \mathbb{R}^n | a_1 x_1 + \cdots + a_n x_n = t\}.
$$

Here, the fuzzy set $\tilde{g}_i(x, \tilde{a}_{i1}, \ldots, \tilde{a}_{i1})$ is denoted as $\tilde{a}_{i1} x_1 \tilde{+} \cdots \tilde{+} \tilde{a}_{in} x_n$, i.e.,

$$
\tilde{g}_i(x, \tilde{a}_{i1}, \ldots, \tilde{a}_{i1}) = \tilde{a}_{i1} x_1 \tilde{+} \cdots \tilde{+} \tilde{a}_{in} x_n
$$

for every $i \in \mathcal{M}$ and for each $x \in \mathbb{R}^n$. The following proposition can be easily derived from the definition.

Proposition 5.23. Let $\tilde{a}_j \in \mathcal{F}_0(\mathbb{R}), x_j \geq 0, j \in \mathcal{N}$. Then $\tilde{a}_1 x_1 \tilde{+} \cdots \tilde{+} \tilde{a}_n x_n$ defined by the extension principle is again a fuzzy quantity.

In (5.31) the value $\tilde{a}_{i1} x_1 \tilde{+} \cdots \tilde{+} \tilde{a}_{in} x_n \in \mathcal{F}_0(\mathbb{R})$ is "compared to" the fuzzy quantity $\tilde{b}_i \in \mathcal{F}_0(\mathbb{R})$ by fuzzy relation $\tilde{R}_i, i \in \mathcal{M}$.

Usually, the fuzzy relations $\tilde{R}_i$ on $\mathbb{R}$ for comparing the left and right sides of the constraints of (5.31) are extensions of a valued relation on $\mathbb{R}$, particularly, the binary inequality relations "$\leq$" or "$\geq$".

If $\tilde{R}_i$ is the T-fuzzy extension of relation $R_i, i \in \mathcal{M}$, then the membership function of the ith constraint is as follows,

$$
\mu_{\tilde{R}_i}(\tilde{a}_{i1} x_1 \tilde{+} \cdots \tilde{+} \tilde{a}_{in} x_n, \tilde{b}_i) = \sup\{T(\mu_{\tilde{a}_{i1} x_1 \tilde{+} \cdots \tilde{+} \tilde{a}_{in} x_n}(u), \mu_{\tilde{b}_i}(v)) | u R_i v\}.
$$

For aggregating fuzzy constraints in FLP problem (5.31), we need some operators with reasonable properties. Such operators should assign to each tuple of elements a unique real number. For this purpose, t-norms or t-conorms can be applied. However, we know some other useful operators generalizing usual t-norms or t-conorms. Clearly, between arbitrary interval $[a, b]$ in $\mathbb{R}$ and the unit interval $[0, 1]$ there exists a one-to-one correspondence. Hence, each result for operators on the interval $[a, b]$ can be transformed into a result for operators on $[0, 1]$ and vice versa. Moreover, the aggregation operators on $[0, 1]$ should be sufficiently general, at least from a theoretical point of view. In many cases, general aggregation operators can be derived from n-ary operations on $[0, 1]$.

Definition 5.24. *An* aggregation operator *G is a sequence $\{G_n\}_{n=1}^{\infty}$ of mappings (called* aggregating mappings*) $G_n : [0, 1]^n \rightarrow [0, 1]$, satisfying the following properties.*

(i) $G_1(x) = x$ for each $x \in [0, 1]$;
(ii) $G_n(x_1, x_2, \ldots, x_n) \leq G_n(y_1, y_2, \ldots, y_n)$, whenever $x_i \leq y_i$ for each $i = 1, 2, \ldots, n$, and every $n = 2, 3, \ldots$;
(iii) $G_n(0, 0, \ldots, 0) = 0$ and $G_n(1, 1, \ldots, 1) = 1$ for every $n = 2, 3, \ldots$.

Condition (i) says that G_1 is a unary identity operation, (ii) means that aggregating mapping G_n is monotone, particularly nondecreasing in all of its arguments x_i, and condition (iii) represents the boundary conditions. Here we have several examples of aggregation operators (see, e.g., [196], [134]):

(1) t-norms and t-conorms;
(2) Usual averages: the arithmetic mean, geometric mean, harmonic mean and root-power mean;
(3) k-order statistic aggregation operators;
(4) Order weighted averaging (OWA) operators;
(5) Sugeno and Choquet integrals.

5.5 Feasible solution

Let us begin with the concept of feasible solution of an FLP problem (5.31).

Definition 5.25. *Let g_i, $i \in \mathcal{M}$, be linear functions defined by (5.28). Let $\mu_{\tilde{a}_{ij}} : \mathbb{R} \rightarrow [0, 1]$ and $\mu_{\tilde{b}_i} : \mathbb{R} \rightarrow [0, 1]$, $i \in \mathcal{M}$, $j \in \mathcal{N}$, be membership functions of fuzzy quantities $\tilde{a}_{ij}$ and $\tilde{b}_i$, respectively. Let $\tilde{R}_i$, $i \in \mathcal{M}$, be fuzzy relations on $\mathbb{R}$. Let G_A be an aggregation operator and T be a t-norm.*

A fuzzy set $\tilde{X}$, the membership function $\mu_{\tilde{X}}$ of which is defined for all $x \in \mathbb{R}^n$ by

$$\mu_{\tilde{X}}(x) = G_A(\mu_{\tilde{R}_1}(\tilde{a}_{11}x_1 \tilde{+} \cdots \tilde{+} \tilde{a}_{1n}x_n, \tilde{b}_1), \cdots, \mu_{\tilde{R}_m}(\tilde{a}_{m1}x_1 \tilde{+} \cdots \tilde{+} \tilde{a}_{mn}x_n, \tilde{b}_m))$$

if $x_j \geq 0$ for all $j \in \mathcal{N}$ and by

$$\mu_{\tilde{X}}(x) = 0 \qquad otherwise, \tag{5.35}$$

is called the feasible solution of the FLP problem (5.31).

For $\alpha \in (0,1]$, a vector $x \in [\tilde{X}]_\alpha$ is called the α-feasible solution of the FLP problem (5.31).

A vector $\bar{x} \in \mathbb{R}^n$ such that $\mu_{\tilde{X}}(\bar{x}) = \mathrm{Hgt}(\tilde{X})$ is called the max-feasible solution.

By the definition the feasible solution $\tilde{X}$ of an FLP problem is a fuzzy set. On the other hand, the α-feasible solution is a vector belonging to the α-cut of the feasible solution $\tilde{X}$ and the same is true for the max-feasible solution, a special α-feasible solution with $\alpha = \mathrm{Hgt}(\tilde{X})$.

Given a feasible solution $\tilde{X}$ and $\alpha \in (0,1]$ (the degree of possibility, feasibility, satisfaction etc.), any vector $x \in \mathbb{R}^n$ satisfying $\mu_{\tilde{X}}(x) \geq \alpha$ is the α-feasible solution of the corresponding FLP problem.

For $i \in \mathcal{M}$, $\tilde{X}_i$ denotes the fuzzy subset of $\mathbb{R}^n$ with the membership function $\mu_{\tilde{X}_i}$ defined for all $x \in \mathbb{R}^n$ as

$$\mu_{\tilde{X}_i}(x) = \mu_{\tilde{R}_i}(\tilde{a}_{i1}x_1 \tilde{+} \ldots \tilde{+} \tilde{a}_{in}x_n, \tilde{b}_i). \tag{5.36}$$

Fuzzy set (5.36) is interpreted as the ith fuzzy constraint. All fuzzy constraints $\tilde{X}_i$ are aggregated into the feasible solution (5.35) by the aggregation operator G_A. Usually, $G_A = \min$ is used for aggregating the constraints; similarly, the t-norm $T = \min$ is used for extending arithmetic operations "$\tilde{+}$".

Clearly, if a_{ij} and b_i are crisp parameters (i.e., crisp fuzzy numbers), then the feasible solution is also crisp. Moreover, if for all $i \in \mathcal{M}$, $\tilde{R}_i$ are T-fuzzy extensions of valued relations R_i and for two collections of fuzzy parameters it holds that $\tilde{a}'_{ij} \subseteq \tilde{a}''_{ij}$ and $\tilde{b}'_i \subseteq \tilde{b}''_i$, then the same holds for the feasible solutions; i.e., $\tilde{X}' \subseteq \tilde{X}''$. See also Proposition 5.30 below.

Now, we derive special formulae that allow for computing an α-feasible solution $x \in [\tilde{X}]_\alpha$ of the FLP problem (5.31). For this purpose, the following notation is useful. Given $\alpha \in (0,1]$, $i \in \mathcal{M}$, $j \in \mathcal{N}$, let $\tilde{a} \in \mathcal{F}_0(\mathbb{R})$. We denote

$$\tilde{a}^{\mathrm{L}}(\alpha) = \inf\{t \in \mathbb{R} | t \in [\tilde{a}]_\alpha\} = \inf[\tilde{a}]_\alpha, \; \tilde{a}^{\mathrm{R}}(\alpha) = \sup\{t | t \in [\tilde{a}]_\alpha\} = \sup[\tilde{a}]_\alpha. \tag{5.37}$$

Theorem 5.26. Let $\tilde{a}_{ij}$ and $\tilde{b}_i$ be fuzzy quantities and $x_j \geq 0$ for all $i \in \mathcal{M}$, $j \in \mathcal{N}$, $\alpha \in (0,1)$. Let $\tilde{\leq}^{\min}$ and $\tilde{\leq}^{\max}$ be fuzzy extensions of the binary relation $\leq$. Then for $i \in \mathcal{M}$ it holds that

(i) $\mu_{\tilde{\leq}^{\min}}(\tilde{a}_{i1}x_1 \tilde{+} \ldots \tilde{+} \tilde{a}_{in}x_n, \tilde{b}_i) \geq \alpha$ if and only if

$$\sum_{j \in \mathcal{N}} \tilde{a}_{ij}^{\mathrm{L}}(\alpha)x_j \leq \tilde{b}_i^{\mathrm{R}}(\alpha), \tag{5.38}$$

(ii) $\mu_{\tilde{\leq}^{\max}}(\tilde{a}_{i1}x_1 \tilde{+} \ldots \tilde{+} \tilde{a}_{in}x_n, \tilde{b}_i) \geq \alpha$ if and only if

$$\sum_{j \in \mathcal{N}} \tilde{a}_{ij}^{R}(1 - \alpha)x_j \leq \tilde{b}_i^{L}(1 - \alpha). \qquad (5.39)$$

Proof. (i) By Proposition 5.23, definition (5.37) and Theorem 5.21 (i), we obtain the required result.

(ii) To prove (i), only normality and compactness of $\tilde{a}_{ij}$ and $\tilde{b}_i$ are utilized; no assumption of convexity has been necessary. Here, in order to apply again Theorem 5.21, (ii), it remains to prove that $\inf[\tilde{a}_{ij}]_\alpha = \inf(\tilde{a}_{ij})_\alpha$, $\sup[\tilde{a}_{ij}]_\alpha = \sup(\tilde{a}_{ij})_\alpha$, $\inf[\tilde{b}_i]_\alpha = \inf(\tilde{b}_i)_\alpha$ and $\sup[\tilde{b}_i]_\alpha = \sup(\tilde{b}_i)_\alpha$. Remember that by (5.37) $\tilde{a}_{ij}^{L}(\alpha) = \inf[\tilde{a}_{ij}]_\alpha$, etc. Evidently, it is sufficient to prove

$$\inf[\tilde{a}]_\alpha = \inf(\tilde{a})_\alpha, \qquad (5.40)$$

$$\sup[\tilde{a}]_\alpha = \sup(\tilde{a})_\alpha, \qquad (5.41)$$

for any fuzzy quantity $\tilde{a} \in \mathcal{F}_0(\mathbb{R})$, i.e., the normal, compact fuzzy subset of $\mathbb{R}$ with the semistrictly quasiconcave membership function. Here we prove only (5.40); identity (5.41) can be proven analogously. Let $\alpha \in (0, 1)$.

1. By definition of the α-cut and strict α-cut we have $(\tilde{a})_\alpha \subseteq [\tilde{a}]_\alpha$. Then $\inf[\tilde{a}]_\alpha \leq \inf(\tilde{a})_\alpha$.

2. Assume $\inf[\tilde{a}]_\alpha < \inf(\tilde{a})_\alpha$. Then there exist x, x_0 such that $\inf[\tilde{a}]_\alpha < x < x_0 < \inf(\tilde{a})_\alpha$ with

$$\mu_{\tilde{a}}(x) = \mu_{\tilde{a}}(x_0) = \alpha. \qquad (5.42)$$

On the other hand, there exists y such that $\inf(\tilde{a})_\alpha < y$ and $\mu_{\tilde{a}}(y) > \alpha$. It follows that $x < x_0 < y$. Then there is $\lambda \in (0, 1)$ with $x_0 = \lambda x + (1 - \lambda)y$. By (5.42) we obtain

$$\mu_{\tilde{a}}(x_0) = \alpha = \min\{\mu_{\tilde{a}}(x), \mu_{\tilde{a}}(y)\}. \qquad (5.43)$$

However, since $\mu_{\tilde{a}}(x) < \mu_{\tilde{a}}(y)$ and $\mu_{\tilde{a}}(x_0) = \mu_{\tilde{a}}(\lambda x + (1 - \lambda)y) > 0$, by semistrict quasiconcavity of $\mu_{\tilde{a}}$ it should be satisfied that

$$\mu_{\tilde{a}}(x_0) > \min\{\mu_{\tilde{a}}(x), \mu_{\tilde{a}}(y)\},$$

a contradiction to (5.43). Consequently, $\inf[\tilde{a}]_\alpha \geq \inf(\tilde{a})_\alpha$.

The rest of the proof follows easily from Proposition 5.23 and Theorem 5.21 (ii). $\qquad \square$

Notice that semistrict quasiconcavity of fuzzy quantities is a property securing validity of the equivalence (ii) in Theorem 5.26, which plays a key role in deriving the duality principle in FLP we deal with later on.

In the following example we apply Theorem 5.26 to a broad and practical class of so-called $(\mathcal{L}, \mathcal{R})$-fuzzy quantities with membership functions given by shifts and contractions of special generator functions.

Example 5.27. Let $l, r \in \mathbb{R}$ with $l \leq r$, let $\gamma, \delta \in [0, +\infty)$ and let $\mathcal{L}, \mathcal{R}$ be non-increasing, upper-semicontinuous, semistrictly quasiconcave functions mapping interval $[0, +\infty)$ into $[0, 1]$; i.e., $\mathcal{L}, \mathcal{R} : [0, +\infty) \to [0, 1]$. Moreover, assume that $\mathcal{L}(0) = \mathcal{R}(0) = 1$ and $\lim_{x \to +\infty} \mathcal{L}(x) = \lim_{x \to +\infty} \mathcal{R}(x) = 0$, for each $x \in \mathbb{R}$,

$$
\mu_A(x) = \begin{cases} \mathcal{L}\left(\frac{l-x}{\gamma}\right) & \text{if } x \in (l - \gamma, l), \ \gamma > 0, \\ 1 & \text{if } x \in [l, r], \\ \mathcal{R}\left(\frac{x-r}{\delta}\right) & \text{if } x \in (r, r + \delta), \ \delta > 0, \\ 0 & \text{otherwise.} \end{cases} \tag{5.44}
$$

We write $A = (l, r, \gamma, \delta)_{\mathcal{LR}}$, the fuzzy quantity A is called an *$(\mathcal{L},\mathcal{R})$-fuzzy interval* and the set of all $(\mathcal{L}, \mathcal{R})$-fuzzy intervals is denoted by $\mathcal{F}_{\mathcal{LR}}(\mathbb{R})$. Observe that $\mathrm{Core}(A) = [l, r]$ and $[A]_\alpha$ is a compact interval for every $\alpha \in (0, 1]$. It is obvious that the class of $(\mathcal{L}, \mathcal{R})$-fuzzy intervals extends the class of crisp closed intervals $[a, b] \subseteq \mathbb{R}$ including the case $a = b$, i.e., crisp numbers. Similarly, if the membership functions of $\tilde{a}_{ij}$ and $\tilde{b}_i$ are given analytically by

$$
\mu_{\tilde{a}_{ij}}(x) = \begin{cases} \mathcal{L}\left(\frac{l_{ij}-x}{\gamma_{ij}}\right) & \text{if } x \in [l_{ij} - \gamma_{ij}, l_{ij}), \ \gamma_{ij} > 0, \\ 1 & \text{if } x \in [l_{ij}, r_{ij}], \\ \mathcal{R}\left(\frac{x-r_{ij}}{\delta_{ij}}\right) & \text{if } x \in (r_{ij}, r_{ij} + \delta_{ij}], \ \delta_{ij} > 0, \\ 0 & \text{otherwise,} \end{cases} \tag{5.45}
$$

and

$$
\mu_{\tilde{b}_j}(x) = \begin{cases} \mathcal{L}\left(\frac{l_i-x}{\gamma_i}\right) & \text{if } x \in [l_i - \gamma_i, l_i), \ \gamma_i > 0, \\ 1 & \text{if } x \in [l_i, r_i], \\ \mathcal{R}\left(\frac{x-r_i}{\delta_i}\right) & \text{if } x \in (r_i, r_i + \delta_i], \ \delta_i > 0, \\ 0 & \text{otherwise,} \end{cases} \tag{5.46}
$$

for each $x \in \mathbb{R}$, $i \in \mathcal{M}$, $j \in \mathcal{N}$, then the values of (5.37) can be computed as

$$
\tilde{a}_{ij}^{\mathrm{L}}(\alpha) = l_{ij} - \gamma_{ij}\mathcal{L}^{(-1)}(\alpha), \qquad \tilde{a}_{ij}^{\mathrm{R}}(\alpha) = r_{ij} + \delta_{ij}\mathcal{R}^{(-1)}(\alpha),
$$
$$
\tilde{b}_i^{\mathrm{L}}(\alpha) = l_i - \gamma_i\mathcal{L}^{(-1)}(\alpha), \qquad \tilde{b}_i^{\mathrm{R}}(\alpha) = r_i + \delta_i\mathcal{R}^{(-1)}(\alpha),
$$

where $\mathcal{L}^{(-1)}$ and $\mathcal{R}^{(-1)}$ are pseudo-inverse functions of $\mathcal{L}$ and $\mathcal{R}$ defined by $\mathcal{L}^{(-1)}(\alpha) = \sup\{x | \mathcal{L}(x) \geq \alpha\}$ and $\mathcal{R}^{(-1)}(\alpha) = \sup\{x | \mathcal{R}(x) \geq \alpha\}$, respectively. Let $G_A = \min$. By Theorem 5.26, the α-cut $[\tilde{X}]_\alpha$ of the feasible solution of (5.31) with $\tilde{R}_i = \precsim^{\min}, i \in \mathcal{M}$, can be obtained by solving the system of inequalities

$$
\sum_{j \in \mathcal{N}} (l_{ij} - \gamma_{ij}\mathcal{L}^{(-1)}(\alpha))x_j \leq r_i + \delta_i\mathcal{R}^{(-1)}(\alpha), i \in \mathcal{M}. \tag{5.47}
$$

On the other hand, the α-cut $[\tilde{X}]_\alpha$ of the feasible solution of (5.31) with $\tilde{R}_i = \precsim^{\max}, i \in \mathcal{M}$, can be obtained by solving the system of inequalities

$$\sum_{j \in \mathcal{N}} (r_{ij} + \delta_{ij} \mathcal{R}^{(-1)}(\alpha)) x_j \le l_i - \gamma_i \mathcal{L}^{(-1)}(\alpha), \; i \in \mathcal{M}. \tag{5.48}$$

Moreover, by (5.47), (5.48), $[\tilde{X}]_\alpha$ is the intersection of a finite number of halfspaces, hence a convex polyhedral set.

5.6 "Optimal" solution

The "optimization", i.e., "maximization" or "minimization", of the objective function requires a special approach, as the set of fuzzy values of the objective function is not linearly ordered. In order to "maximize" the objective function we introduce a suitable concept of "optimal solution". It is done by two distinct approaches, namely: (1) satisficing solution or (2) α-efficient solution.

5.6.1 Satisficing solution

We assume the existence of an exogenously given goal $\tilde{d} \in \mathcal{F}(\mathbb{R})$. The fuzzy value $\tilde{d}$ is compared to fuzzy values $\tilde{c}_1 x_1 \tilde{+} \cdots \tilde{+} \tilde{c}_n x_n$ of the objective function by a given fuzzy relation $\tilde{R}_0$. In this way the fuzzy objective function is treated as another constraint

$$(\tilde{c}_1 x_1 \tilde{+} \cdots \tilde{+} \tilde{c}_n x_n) \tilde{R}_0 \; \tilde{d}.$$

The satisficing solution is then obtained by a modification of the definition of the feasible solution.

Definition 5.28. *Let f, g_i be linear functions defined by (5.27), (5.28). Let $\mu_{\tilde{c}_j} : \mathbb{R} \to [0,1]$, $\mu_{\tilde{a}_{ij}} : \mathbb{R} \to [0,1]$ and let $\mu_{\tilde{b}_i} : \mathbb{R} \to [0,1]$, $i \in \mathcal{M}$, $j \in \mathcal{N}$, be membership functions of fuzzy quantities $\tilde{c}_j$, $\tilde{a}_{ij}$ and $\tilde{b}_i$, respectively. Moreover, let $\tilde{d} \in \mathcal{F}_I(\mathbb{R})$ be a fuzzy interval, called the fuzzy goal. Let $\tilde{R}_i$, $i \in \{0\} \cup \mathcal{M}$, be fuzzy relations on $\mathbb{R}$ and T be a t-norm and G and G_A be aggregation operators.*

A fuzzy set $\tilde{X}^*$ with the membership function $\mu_{\tilde{X}^*}$ defined for all $x \in \mathbb{R}^n$ by

$$\mu_{\tilde{X}^*}(x) = G_A(\mu_{\tilde{R}_0}(\tilde{c}_1 x_1 \tilde{+} \cdots \tilde{+} \tilde{c}_n x_n, \tilde{d}), \mu_{\tilde{X}}(x)), \tag{5.49}$$

where $\mu_{\tilde{X}}(x)$ is the membership function of the feasible solution, is called the satisficing solution of the FLP problem (5.31).

For $\alpha \in (0,1]$ a vector $x \in [\tilde{X}^*]_\alpha$ is called the α-satisficing solution of the FLP problem (5.31).

A vector $x^* \in \mathbb{R}^n$ with the property

$$\mu_{\tilde{X}^*}(x^*) = \mathrm{Hgt}(\tilde{X}^*) \tag{5.50}$$

is called the max-satisficing solution.

By Definition 5.28 any satisficing solution of the FLP problem is a fuzzy set. On the other hand, the α-satisficing solution belongs to the α-cut $[\tilde{X}^*]_\alpha$. Likewise, the max-satisficing solution is an α-satisficing solution with $\alpha = \mathrm{Hgt}(\tilde{X}^*)$.

The t-norm T is used for extending arithmetic operations, the aggregation operator G for joining the individual constraints into the feasible solution and G_A is applied for aggregating the fuzzy set of the feasible solution and fuzzy set of the objective $\tilde{X}_0$ defined by the membership function

$$\mu_{\tilde{X}_0}(x) = \mu_{\tilde{R}_0}(\tilde{c}_1 x_1 \tilde{+} \cdots \tilde{+} \tilde{c}_n x_n, \tilde{d}), \tag{5.51}$$

for all $x \in \mathbb{R}^n$. The membership function of optimal solution $\tilde{X}^*$ is defined for all $x \in \mathbb{R}^n$ by

$$\mu_{\tilde{X}^*}(x) = G_A(\mu_{\tilde{X}_0}(x), \mu_{\tilde{X}}(x)). \tag{5.52}$$

If (5.31) is a maximization problem "the higher value is better," then the membership function $\mu_{\tilde{d}}$ of the fuzzy goal $\tilde{d}$ is supposed to be increasing or nondecreasing. If (5.31) is a minimization problem "the lower value is better," then the membership function $\mu_{\tilde{d}}$ of $\tilde{d}$ is decreasing or nonincreasing. The fuzzy relation $\tilde{R}_0$ for comparing $\tilde{c}_1 x_1 \tilde{+} \cdots \tilde{+} \tilde{c}_n x_n$ and $\tilde{d}$ is supposed to be a fuzzy extension of $\geq$ or $\leq$.

Formally, Definitions 5.25 and 5.28 are similar. In other words, the concept of the feasible solution is similar to the concept of optimal solution. Therefore, we can take advantage of the properties of the feasible solution studied in the preceding section.

Observe that in the case of crisp parameters c_j, a_{ij} and b_i, the set of all max-optimal solutions given by (5.50) coincides with the set of all optimal solutions of the classical LP problem. We have the following result.

Proposition 5.29. *Let c_j, a_{ij}, $b_i \in \mathbb{R}$ be crisp fuzzy numbers for all $i \in \mathcal{M}$, $j \in \mathcal{N}$. Let $\tilde{d} \in \mathcal{F}(\mathbb{R})$ be a fuzzy goal with a strictly increasing membership function $\mu_{\tilde{d}}$. Let for $i \in \mathcal{M}$, $\tilde{R}_i$ be a fuzzy extension of relation "$\leq$" on $\mathbb{R}$, and $\tilde{R}_0$ be a T-fuzzy extension of relation "$\geq$". Let T, G and G_A be t-norms.*

Then the set of all max-satisficing solutions of (5.31) coincides with the set of all optimal solutions X^* of LP problem (5.29).

Proof. Clearly, a feasible solution $\tilde{X}$ of (5.31) is crisp; i.e., $\mu_{\tilde{X}}(x) = \chi_X(x)$ for all $x \in \mathbb{R}^n$, where X is the set of all feasible solutions (5.30) of crisp LP problem (5.29). Moreover, by (5.51), for crisp $c \in \mathbb{R}^n$ we obtain

$$\mu_{\tilde{X}_0}(x) = \mu_{\tilde{R}_0}(f(x, c), \tilde{d}) = \mu_{\tilde{d}}(c_1 x_1 + \cdots + c_n x_n).$$

Substituting into (5.52) we get

$$\mu_{\tilde{X}^*}(x) = G_A(\mu_{\tilde{d}}(f(x, c)), \chi_X(x)) = \begin{cases} \mu_{\tilde{d}}(c_1 x_1 + \cdots + c_n x_n) & \text{if } x \in X, \\ 0 & \text{otherwise.} \end{cases}$$

Since $\mu_{\tilde{d}}$ is strictly increasing, it follows that

$$\mu_{\tilde{X}^*}(x^*) = \mathrm{Hgt}(\tilde{X}^*)$$

if and only if

$$\mu_{\tilde{X}^*}(x^*) = \sup\{\mu_{\tilde{d}}(c_1 x_1 + \cdots + c_n x_n) \mid x \in X\},$$

which is the desired result. $\qquad\qquad\qquad\qquad\qquad\qquad\square$

Proposition 5.30. *Let $\tilde{c}'_j$, $\tilde{a}'_{ij}$ and $\tilde{b}'_i$ and $\tilde{c}''_j$, $\tilde{a}''_{ij}$ and $\tilde{b}''_i$ be two collections of fuzzy quantities, parameters of FLP problem (5.31), $i \in \mathcal{M}$, $j \in \mathcal{N}$. Let T, G, G_A be t-norms. Let $\tilde{R}_i$, $i \in \{0\} \cup \mathcal{M}$, be T-fuzzy extensions of valued relations R_i on $\mathbb{R}$, and $\tilde{d} \in \mathcal{F}_I(\mathbb{R})$ be a fuzzy goal.*

If $\tilde{X}^{*\prime}$ is the satisficing solution of FLP problem (5.31) with the parameters $\tilde{c}'_j$, $\tilde{a}'_{ij}$ and $\tilde{b}'_i$, $\tilde{X}^{*\prime\prime}$ is the satisficing solution of the FLP problem with the parameters $\tilde{c}''_j$, $\tilde{a}''_{ij}$ and $\tilde{b}''_i$ such that for all $i \in \mathcal{M}$, $j \in \mathcal{N}$,

$$\tilde{c}'_j \subseteq \tilde{c}''_j, \ \tilde{a}'_{ij} \subseteq \tilde{a}''_{ij} \text{ and } \tilde{b}'_i \subseteq \tilde{b}''_i,$$

then it holds

$$\tilde{X}^{*\prime} \subseteq \tilde{X}^{*\prime\prime}.$$

Proof. First, we show that $\tilde{X}' \subseteq \tilde{X}''$. Let $x \in \mathbb{R}^n$, $i \in \mathcal{M}$. We must show that

$$\tilde{a}'_{i1} x_1 \tilde{+} \cdots \tilde{+} \tilde{a}'_{in} x_n \subseteq \tilde{a}''_{i1} x_1 \tilde{+} \cdots \tilde{+} \tilde{a}''_{in} x_n.$$

For each $u \in \mathbb{R}$ we get

$$\mu_{\tilde{a}'_{i1} x_1 \tilde{+} \cdots \tilde{+} \tilde{a}'_{in} x_n}(u)$$
$$= \sup\{T(\mu_{\tilde{a}'_{i1}}(a_1), \ldots, \mu_{\tilde{a}'_{in}}(a_n)) \mid a_{i1} x_1 + \cdots + a_{in} x_n = u\}$$
$$\leq \sup\{T(\mu_{\tilde{a}''_{i1}}(a_1), \ldots, \mu_{\tilde{a}''_{in}}(a_n)) \mid a_{i1} x_1 + \cdots + a_{in} x_n = u\}$$
$$= \mu_{\tilde{a}''_{i1} x_1 \tilde{+} \cdots \tilde{+} \tilde{a}''_{in} x_n}(u).$$

Now, as $\tilde{b}'_i \subseteq \tilde{b}''_i$, using monotonicity of T-fuzzy extension $\tilde{R}_i$ of R_i, we obtain

$$\mu_{\tilde{R}_i}(\tilde{a}'_{i1} x_1 \tilde{+} \cdots \tilde{+} \tilde{a}'_{in} x_n, \tilde{b}'_i) \leq \mu_{\tilde{R}_i}(\tilde{a}''_{i1} x_1 \tilde{+} \cdots \tilde{+} \tilde{a}''_{in} x_n, \tilde{b}''_i).$$

Then, applying again monotonicity of G in (5.35), we obtain $\tilde{X}' \subseteq \tilde{X}''$.

It remains to show that $\tilde{X}'_0 \subseteq \tilde{X}''_0$, where

$$\mu_{\tilde{X}'_0}(x) = \mu_{\tilde{R}_0}(\tilde{f}(x, \tilde{c}'), \tilde{d}), \ \mu_{\tilde{X}''_0}(x) = \mu_{\tilde{R}_0}(\tilde{f}(x, \tilde{c}''), \tilde{d}).$$

We show that $\tilde{f}(x, \tilde{c}') \subseteq \tilde{f}(x, \tilde{c}'')$. Since for all $j \in \mathcal{N}$, $\mu_{\tilde{c}'_j}(c) \leq \mu_{\tilde{c}''_j}(c)$ for all $c \in \mathbb{R}$, by (5.32) we obtain for all $u \in \mathbb{R}$,

$$\mu_{\tilde{c}'_1 x_1 \tilde{+} \cdots \tilde{+} \tilde{c}'_n x_n}(u)$$
$$= \sup\{T(\mu_{\tilde{c}'_1}(c_1), \ldots, \mu_{\tilde{c}'_n}(c_n)) \mid c_1 x_1 + \cdots + c_n x_n = u\}$$
$$\leq \sup\{T(\mu_{\tilde{c}''_1}(c_1), \ldots, \mu_{\tilde{c}''_n}(c_n)) \mid c_1 x_1 + \cdots + c_n x_n = u\}$$
$$= \mu_{\tilde{c}''_1 x_1 \tilde{+} \cdots \tilde{+} \tilde{c}''_n x_n}(u).$$

Using monotonicity of $\tilde{R}_0$, we have

$$\mu_{\tilde{R}_0}(\tilde{c}'_1 x_1 \tilde{+} \cdots \tilde{+} \tilde{c}'_n x_n, \tilde{d}) \leq \mu_{\tilde{R}_0}(\tilde{c}''_1 x_1 \tilde{+} \cdots \tilde{+} \tilde{c}''_n x_n, \tilde{d}).$$

Finally, applying monotonicity of G_A in (5.52), we obtain $\tilde{X}^{*\prime} \subseteq \tilde{X}^{*\prime\prime}$. □

Further on, we extend Theorem 5.26 to the case of the satisficing solution of an FLP problem. For this purpose we introduce the following notation. Given $\alpha \in (0, 1]$, $j \in \mathcal{N}$, let

$$\tilde{c}^L_j(\alpha) = \inf\{c \mid c \in [\tilde{c}_j]_\alpha\},$$
$$\tilde{c}^R_j(\alpha) = \sup\{c \mid c \in [\tilde{c}_j]_\alpha\},$$
$$\tilde{d}^L(\alpha) = \inf\{d \mid d \in [\tilde{d}]_\alpha\},$$
$$\tilde{d}^R(\alpha) = \sup\{d \mid d \in [\tilde{d}]_\alpha\}.$$

Theorem 5.31. *Let $\tilde{c}_j$, $\tilde{a}_{ij}$ and $\tilde{b}_i$ be fuzzy quantities, $i \in \mathcal{M}$, $j \in \mathcal{N}$. Let $\tilde{d} \in \mathcal{F}(\mathbb{R})$ be a fuzzy goal with the membership function $\mu_{\tilde{d}}$ satisfying the following conditions,*

$$\begin{aligned} &\mu_{\tilde{d}} \text{ is upper semicontinuous,}\\ &\mu_{\tilde{d}} \text{ is strictly increasing,} \qquad\qquad (5.53)\\ &\lim\nolimits_{t \to -\infty} \mu_{\tilde{d}}(t) = 0. \end{aligned}$$

For $i \in \mathcal{M}$, let $\tilde{R}_i$ be the T-fuzzy extension of the binary relation $\leq$ on $\mathbb{R}$, and $\tilde{R}_0$ be the T-fuzzy extension of the binary relation $\geq$ on $\mathbb{R}$. Let $T = G = G_A = \min$. Let $\tilde{X}^*$ be a satisficing solution of FLP problem (5.31) and let $\alpha \in (0, 1)$. A vector $x = (x_1, \ldots, x_n) \geq 0$ belongs to $[\tilde{X}^*]_\alpha$ if and only if

$$\sum_{j=1}^n \tilde{c}^R_j(\alpha) x_j \geq \tilde{d}^L(\alpha), \qquad\qquad (5.54)$$

$$\sum_{j=1}^n \tilde{a}^L_{ij}(\alpha) x_j \leq \tilde{b}^R_i(\alpha), \quad i \in \mathcal{M}. \qquad\qquad (5.55)$$

The proof is omitted; it is analogous to the proof of Theorem 5.26, part (i), with a simple modification: instead of compactness of $\tilde{d}$, we assume (5.53).

If the membership functions of the fuzzy parameters $\tilde{c}_j$, $\tilde{a}_{ij}$ and $\tilde{b}_i$ can be formulated in an explicit form, e.g., as $(\mathcal{L}, \mathcal{R})$-fuzzy quantities (see (5.46)), then we can find a max-satisficing solution as the optimal solution of some associated classical optimization problem.

Proposition 5.32. *Let*

$$\mu_{\tilde{X}_0}(x) = \mu_{\tilde{R}_0}(\tilde{c}_1 x_1 \; \tilde{+} \cdots \tilde{+} \; \tilde{c}_n x_n, \tilde{d})$$

be the membership function of the fuzzy objective and let

$$\mu_{\tilde{X}_i}(x) = \mu_{\tilde{R}_i}(\tilde{a}_{i1} x_1 + \cdots + \tilde{a}_{in} x_n, \tilde{b}_i), \quad i \in \mathcal{M},$$

be the membership functions of the fuzzy constraints, $x = (x_1, \ldots, x_n) \in \mathbb{R}^n$. Let $T = G = G_A = \min$ and assume that (5.53) holds for fuzzy goal $\tilde{d}$. Then the vector $(t^*, x^*) \in \mathbb{R}^{n+1}$ is an optimal solution of the optimization problem

$$\begin{aligned} &maximize\ t \\ &subject\ to\ \mu_{\tilde{X}_i}(x) \geq t, \quad i \in \{0\} \cup \mathcal{M}, \quad\quad (5.56) \\ &\quad\quad x_j \geq 0, \quad j \in \mathcal{N} \end{aligned}$$

if and only if $x^* \in \mathbb{R}^n$ is a max-satisficing solution of FLP problem (5.31).

Proof. Let $(t^*, x^*) \in \mathbb{R}^{n+1}$ be an optimal solution of problem (5.56). By (5.49) and (5.50) we obtain

$$\mu_{\tilde{X}^*}(x^*) = \sup\{\min\{\mu_{\tilde{X}_0}(x), \mu_{\tilde{X}_i}(x)\} | x \in \mathbb{R}^n\} = \mathrm{Hgt}(\tilde{X}^*).$$

Hence, x^* is a max-satisficing solution.

The proof of the converse statement follows from Definition 5.28.   □

5.6.2 α-efficient solution

Now, let $\tilde{a}$ and $\tilde{b}$ be fuzzy quantities and $\tilde{R}$ be a fuzzy relation on $\mathbb{R}$, $\alpha \in (0, 1]$. We write

$$\tilde{a} \precsim_\alpha^{\tilde{R}} \tilde{b}, \text{ if } \mu_{\tilde{R}}(\tilde{a}, \tilde{b}) \geq \alpha. \quad\quad (5.57)$$

We also write

$$\tilde{a} \prec_\alpha^{\tilde{R}} \tilde{b}, \text{ if } \tilde{a} \precsim_\alpha^{\tilde{R}} \tilde{b} \text{ and } \mu_{\tilde{R}}(\tilde{b}, \tilde{a}) < \alpha. \quad\quad (5.58)$$

Notice that $\precsim_\alpha^{\tilde{R}}$ is a binary relation on the set of all fuzzy quantities $\mathcal{F}_0(\mathbb{R})$. If $\tilde{a}$ and $\tilde{b}$ are crisp fuzzy numbers corresponding to real numbers a and b, respectively, and $\tilde{R}$ is a fuzzy extension of relation $\leq$, then $\tilde{a} \precsim_\alpha^{\tilde{R}} \tilde{b}$ if and only if $a \leq b$.

Now, modifying the well-known concept of the efficient (nondominated) solution of the LP problem we define "maximization" (or "minimization") of the objective function of the FLP problem (5.31).

Definition 5.33. *Let $\tilde{c}_j$, $\tilde{a}_{ij}$ and $\tilde{b}_i$, $i \in \mathcal{M}$, $j \in \mathcal{N}$, be fuzzy quantities on $\mathbb{R}$. Let $\tilde{R}_i$, $i \in 0, 1, 2, \ldots, m$, be fuzzy relations on $\mathbb{R}$ and $\alpha \in (0, 1]$. Let $x = (x_1, \ldots, x_n)^T$ be an α-feasible solution of (5.31) and denote $\tilde{c}^T x =$*

$\tilde{c}_1 x_1 \,\tilde{+}\, \ldots \,\tilde{+}\, \tilde{c}_n x_n$. The vector $x \in \mathbb{R}^n$ is an α-efficient solution of (5.31) with maximization of the objective function if there is no $x' \in [\tilde{X}]_\alpha$ such that $\tilde{c}^T x \prec_\alpha^{\tilde{R}_0} \tilde{c}^T x'$. Similarly, the vector x is an α-efficient solution of (5.31) with minimization of the objective function if there is no $x' \in [\tilde{X}]_\alpha$ such that $\tilde{c}^T x' \prec_\alpha^{\tilde{R}_0} \tilde{c}^T x$.

Notice that any α-efficient solution of the FLP problem is an α-feasible solution of the FLP problem with some additional property. If all coefficients of FLP problem (5.31) are crisp, then the α-efficient solution of the FLP problem is equivalent to the classical optimal solution of the corresponding LP problem.

In the following theorem we show some necessary and sufficient conditions for an α-efficient solution of (5.31) in the case of special fuzzy extensions of the binary relation $\le$.

Theorem 5.34. Let $\tilde{c}_j$, $\tilde{a}_{ij}$ and $\tilde{b}_i$, $i \in \mathcal{M}$, $j \in \mathcal{N}$, be fuzzy quantities, $\alpha \in (0,1)$.

(i) Let $\tilde{R}_i = \tilde{\le}^{\min}$, i.e., $\tilde{R}_i$ be a fuzzy extension of the binary relation $\le$ on $\mathbb{R}$ defined by (5.19), (5.20) for all $i \in 0, 1, 2, \ldots, m$. Let $x^* = (x_1^*, \ldots, x_n^*)^T$, $x_j^* \ge 0$, $j \in \mathcal{N}$, be an α-feasible solution of (5.31). Then the vector $x^* \in \mathbb{R}^n$ is an α-efficient solution of (5.31) with maximization of the objective function if and only if x^* is an optimal solution of the following LP problem,

$$\text{maximize } \tilde{c}_1^R(\alpha) x_1 + \cdots + \tilde{c}_n^R(\alpha) x_n$$
$$\text{subject to } \tilde{a}_{i1}^L(\alpha) x_1 + \cdots + \tilde{a}_{in}^L(\alpha) x_n \le \tilde{b}_i^R(\alpha), \quad i \in \mathcal{M}, \qquad (5.59)$$
$$x_j \ge 0, \quad j \in \mathcal{N}.$$

(ii) Let $\tilde{R}_0 = \tilde{\le}^{\min}$, $\tilde{R}_i = \tilde{\le}^{\max}$, $i \in 1, 2, \ldots, m$. Let $x^* = (x_1^*, \ldots, x_n^*)^T$, $x_j^* \ge 0$, $j \in \mathcal{N}$, be an α-feasible solution of (5.31). Then the vector $x^* \in \mathbb{R}^n$ is an α-efficient solution of (5.31) with maximization of the objective function if and only if x^* is an optimal solution of the following LP problem,

$$\text{maximize } \tilde{c}_1^R(\alpha) x_1 + \cdots + \tilde{c}_n^R(\alpha) x_n$$
$$\text{subject to } \tilde{a}_{i1}^R(\alpha) x_1 + \cdots + \tilde{a}_{in}^R(\alpha) x_n \le \tilde{b}_i^L(\alpha), \quad i \in \mathcal{M}, \qquad (5.60)$$
$$x_j \ge 0, \quad j \in \mathcal{N}.$$

(iii) Let $\tilde{R}_0 = \tilde{\le}^{\max}$, $\tilde{R}_i = \tilde{\le}^{\min}$, $i \in 1, 2, \ldots, m$. Let $x^* = (x_1^*, \ldots, x_n^*)^T$, $x_j^* \ge 0$, $j \in \mathcal{N}$, be an α-feasible solution of (5.31). Then the vector $x^* \in \mathbb{R}^n$ is an α-efficient solution of (5.31) with maximization of the objective function if and only if x^* is an optimal solution of the following LP problem,

$$\text{maximize } \tilde{c}_1^L(\alpha) x_1 + \cdots + \tilde{c}_n^L(\alpha) x_n$$
$$\text{subject to } \tilde{a}_{i1}^L(\alpha) x_1 + \cdots + \tilde{a}_{in}^L(\alpha) x_n \le \tilde{b}_i^R(\alpha), \quad i \in \mathcal{M}, \qquad (5.61)$$
$$x_j \ge 0, \quad j \in \mathcal{N}.$$

(iv) Let $\tilde{R}_i = \preceq^{\max}$, $i \in 0, 1, 2, \ldots, m$. Let $x^* = (x_1^*, \ldots, x_n^*)^T$, $x_j^* \geq 0$, $j \in \mathcal{N}$, be an α-feasible solution of (5.31). Then the vector $x^* \in \mathbb{R}^n$ is an α-efficient solution of (5.31) with maximization of the objective function if and only if x^* is an optimal solution of the following LP problem,

$$\text{maximize } \tilde{c}_1^L(\alpha)x_1 + \cdots + \tilde{c}_n^L(\alpha)x_n$$
$$\text{subject to } \tilde{a}_{i1}^R(\alpha)x_1 + \cdots + \tilde{a}_{in}^R(\alpha)x_n \leq \tilde{b}_i^L(\alpha), \quad i \in \mathcal{M}, \tag{5.62}$$
$$x_j \geq 0, \quad j \in \mathcal{N}.$$

Proof. First, we prove part (i). Let $x^* = (x_1^*, \ldots, x_n^*)^T$, $x_j^* \geq 0$, $j \in \mathcal{N}$, be an α-efficient solution of (5.31) with maximization of the objective function. By Theorem 5.26 (i), formulae (5.57), (5.58) and Definition 5.33, x^* is an optimal solution of the LP problem (5.59). On the other hand, if x^* is an optimal solution of the LP problem (5.59), then by Theorem 5.26, x^* is an α-feasible solution of the FLP problem (5.31). It is evident that (5.57) and also (5.58) are satisfied. Hence, x^* is an α-efficient solution of (5.31) with maximization of the objective function.

The proof of the other parts is analogous. Here we appropriately use Theorem 5.26 (ii). □

In the following section we investigate duality, a fundamental concept of linear optimization. Again we distinguish the above-mentioned two approaches to "optimality" in FLP.

5.7 Duality

In this section we generalize the well-known concept of duality in LP for FLP problems. Some results of this section can also be found in [135]. We derive some weak and strong duality theorems that extend the known results for LP problems.

Consider the following FLP problem

$$\text{"maximize" } \tilde{c}_1 x_1 \tilde{+} \cdots \tilde{+} \tilde{c}_n x_n$$
$$\text{subject to } (\tilde{a}_{i1}x_1 \tilde{+} \cdots \tilde{+} \tilde{a}_{in}x_n)\tilde{R}\tilde{b}_i, \quad i \in \mathcal{M}, \tag{5.63}$$
$$x_j \geq 0, \quad j \in \mathcal{N},$$

where $\tilde{c}_j$, $\tilde{a}_{ij}$ and $\tilde{b}_i$ are normal fuzzy quantities with membership functions $\mu_{\tilde{c}_j} : \mathbb{R} \to [0, 1]$, $\mu_{\tilde{a}_{ij}} : \mathbb{R} \to [0, 1]$ and $\mu_{\tilde{b}_i} : \mathbb{R} \to [0, 1]$, $i \in \mathcal{M}, j \in \mathcal{N}$.

Let $\Phi : \mathcal{F}(\mathbb{R} \times \mathbb{R}) \to \mathcal{F}(\mathcal{F}(\mathbb{R}) \times \mathcal{F}(\mathbb{R}))$ be a mapping and $\Psi : \mathcal{F}(\mathbb{R} \times \mathbb{R}) \to \mathcal{F}(\mathcal{F}(\mathbb{R}) \times \mathcal{F}(\mathbb{R}))$ be the dual mapping to mapping Φ. Let R be a valued relation on $\mathbb{R}$ and let $\tilde{R} = \Phi(R)$ and $\tilde{R}^D = \Psi(R)$; see Definition 5.15. Then $\tilde{R}$ and $\tilde{R}^D$ are dual fuzzy relations.

FLP problem (5.63) is called the *primal FLP problem (P)*.

The *dual FLP problem (D)* is defined as

$$\text{"minimize"} \quad \tilde{b}_1 y_1 \tilde{+} \cdots \tilde{+} \tilde{b}_m y_m$$
$$\text{subject to} \quad \tilde{c}_j \tilde{R}^D (\tilde{a}_{1j} y_1 \tilde{+} \cdots \tilde{+} \tilde{a}_{mj} y_m), \quad j \in \mathcal{N}, \tag{5.64}$$
$$y_i \geq 0, \quad i \in \mathcal{M}.$$

The pair of FLP problems (5.63) and (5.64) is called the *primal–dual pair of FLP problems*.

Let R be the binary operation $\leq$, let $T = \min$ and $S = \max$. Let $\tilde{\leq}^{\min}$ and $\tilde{\leq}^{\max}$ be fuzzy extensions defined by (5.19) and (5.20), respectively. Since T is the dual t-norm to S, by Definition 5.15, $\tilde{\leq}^{\max}$ is the dual fuzzy relation to $\tilde{\leq}^{\min}$. We obtain the primal–dual pair of FLP problems as follows,
(P):

$$\text{"maximize"} \quad \tilde{c}_1 x_1 \tilde{+} \cdots \tilde{+} \tilde{c}_n x_n$$
$$\text{subject to} \quad \tilde{a}_{i1} x_1 \tilde{+} \cdots \tilde{+} \tilde{a}_{in} x_n \tilde{\leq}^{\min} \tilde{b}_i, \quad i \in \mathcal{M}, \tag{5.65}$$
$$x_j \geq 0, \quad j \in \mathcal{N}.$$

(D):

$$\text{"minimize"} \quad \tilde{b}_1 y_1 \tilde{+} \cdots \tilde{+} \tilde{b}_m y_m$$
$$\text{subject to} \quad \tilde{c}_j \tilde{\leq}^{\max} \tilde{a}_{1j} y_1 \tilde{+} \cdots \tilde{+} \tilde{a}_{mj} y_m, \quad j \in \mathcal{N}, \tag{5.66}$$
$$y_i \geq 0, \quad i \in \mathcal{M}.$$

Let the feasible solution of the primal FLP problem (P) be denoted by $\tilde{X}$ and the feasible solution of the dual FLP problem (D) by $\tilde{Y}$. Clearly, $\tilde{X}$ is a fuzzy subset of $\mathbb{R}^n$; $\tilde{Y}$ is a fuzzy subset of $\mathbb{R}^m$.

Notice that in the crisp case, i.e., when the parameters $\tilde{c}_j$, $\tilde{a}_{ij}$ and $\tilde{b}_i$ are crisp fuzzy numbers, by Theorem 5.26 the relations $\tilde{\leq}^{\min}$ and $\tilde{\leq}^{\max}$ coincide with $\leq$; hence (P) and (D) are a primal–dual pair of LP problems in the classical sense. The following proposition is a useful modification of Theorem 5.26.

Proposition 5.35. *Let $\tilde{c}_j$ and $\tilde{a}_{ij}$ be fuzzy quantities and let $y_i \geq 0$ for all $i \in \mathcal{M}$, $j \in \mathcal{N}$, $\alpha \in (0,1)$. Let $\tilde{\geq}^{\max}$ be a fuzzy extension of the binary relation $\geq$ defined by Definition 5.20. Then for $j \in \mathcal{N}$ it holds*

$$\mu_{\tilde{\geq}^{\max}} (\tilde{a}_{1j} y_1 \tilde{+} \cdots \tilde{+} \tilde{a}_{mj} y_m, \tilde{c}_j) \geq 1 - \alpha \text{ if and only if}$$

$$\sum_{i \in \mathcal{M}} \tilde{a}_{ij}^{\mathrm{L}}(\alpha) y_i \geq \tilde{c}_j^{\mathrm{R}}(\alpha). \tag{5.67}$$

The proof of Proposition 5.35 follows easily from Theorem 5.26. In the following theorem we prove the weak form of the duality theorem for FLP problems.

Theorem 5.36. First Weak Duality Theorem. *Let $\tilde{c}_j$, $\tilde{a}_{ij}$ and $\tilde{b}_i$ be fuzzy quantities for all $i \in \mathcal{M}$ and $j \in \mathcal{N}$. Let $A = T_M = \min$, $S = S_M = \max$ and $\alpha \in (0,1)$. Let $\tilde{X}$ be a feasible solution of FLP problem (5.63) and $\tilde{Y}$ be a feasible solution of FLP problem (5.64).*

If a vector $x = (x_1, \ldots, x_n)^T \geq 0$ belongs to $[\tilde{X}]_\alpha$ and $y = (y_1, \ldots, y_m)^T \geq 0$ belongs to $[\tilde{Y}]_{1-\alpha}$, then

$$\sum_{j \in \mathcal{N}} \tilde{c}_j^{\mathrm{R}}(\alpha) x_j \leq \sum_{i \in \mathcal{M}} \tilde{b}_i^{\mathrm{R}}(\alpha) y_i. \tag{5.68}$$

Proof. Let $x \in [\tilde{X}]_\alpha$ and $y \in [\tilde{Y}]_{1-\alpha}$, $x_j \geq 0, y_i \geq 0$ for all $i \in \mathcal{M}, j \in \mathcal{N}$. Then by Proposition 5.35 and by multiplying both sides by x_j and summing up we obtain

$$\sum_{j \in \mathcal{N}} \sum_{i \in \mathcal{M}} \tilde{a}_{ij}^{\mathrm{L}}(\alpha) y_i x_j \geq \sum_{j \in \mathcal{N}} \tilde{c}_j^{\mathrm{R}}(\alpha) x_j. \tag{5.69}$$

Similarly, by Theorem 5.26 we obtain

$$\sum_{j \in \mathcal{N}} \sum_{i \in \mathcal{M}} \tilde{a}_{ij}^{\mathrm{L}}(\alpha) x_j y_i \leq \sum_{i \in \mathcal{M}} \tilde{b}_i^{\mathrm{R}}(\alpha) y_i. \tag{5.70}$$

Combining inequalities (5.69) and (5.70), we obtain

$$\sum_{j \in \mathcal{N}} \tilde{c}_j^{\mathrm{R}}(\alpha) x_j \leq \sum_{j \in \mathcal{N}} \sum_{i \in \mathcal{M}} \tilde{a}_{ij}^{\mathrm{L}}(\alpha) x_j y_i \leq \sum_{i \in \mathcal{M}} \tilde{b}_i^{\mathrm{R}}(\alpha) y_i,$$

which is the desired result. □

Theorem 5.37. Second Weak Duality Theorem. *Let $\tilde{c}_j$, $\tilde{a}_{ij}$ and $\tilde{b}_i$ be fuzzy quantities for all $i \in \mathcal{M}$ and $j \in \mathcal{N}$. Let $A = T_M = \min$, $S = S_M = \max$ and $\alpha \in (0,1)$. Let $\tilde{X}$ be a feasible solution of FLP problem (5.63) and $\tilde{Y}$ be a feasible solution of FLP problem (5.64).*

If for some $x = (x_1, \ldots, x_n)^T \geq 0$ belonging to $[\tilde{X}]_\alpha$ and $y = (y_1, \ldots, y_m)^T \geq 0$ belonging to $[\tilde{Y}]_{1-\alpha}$ it holds

$$\sum_{j \in \mathcal{N}} \tilde{c}_j^{\mathrm{R}}(\alpha) x_j = \sum_{i \in \mathcal{M}} \tilde{b}_i^{\mathrm{R}}(\alpha) y_i, \tag{5.71}$$

then x is an α-efficient solutions of FLP problem (P) and y is a $(1 - \alpha)$-efficient solution of FLP problem (D).

Proof. Let $x \in [\tilde{X}]_\alpha$ and $y \in [\tilde{Y}]_{1-\alpha}$. Then by Proposition 5.35, inequality (5.68) is satisfied, and, moreover, equality (5.71) holds. Suppose that $x \in [\tilde{X}]_\alpha$ is not an α-efficient solution of FLP problem (P). Then there exists $x' \in [\tilde{X}]_\alpha$ such that $\tilde{c}^T x \prec_\alpha \tilde{c}^T x'$. However, by Definition 5.33, (5.57) and (5.58) we obtain

$$\sum_{j \in \mathcal{N}} \tilde{c}_j^{\mathrm{R}}(\alpha) x_j < \sum_{j \in \mathcal{N}} \tilde{c}_j^{\mathrm{R}}(\alpha) x_j',$$

a contradiction to (5.68). Hence, x is an α-efficient solutions of FLP problem (P).

The second part of the proposition, saying that y is a $(1 - \alpha)$-efficient solutions of FLP problem (D) can be proven analogously.   □

Remarks

1. In the crisp case, Theorems 5.36 and 5.37 are standard LP weak duality theorems.

2. The result of the first weak duality theorem is independent of the "maximization" or "minimization" approach.

3. By analogy we can easily formulate the primal–dual pair of FLP problems interchanging the fuzzy relations $\lesssim^{\min}$ and $\lesssim^{\max}$ in the objective functions and/or constraints of (5.63) and (5.64). Then the weak duality theorems should be appropriately modified.

4. Let $\alpha \geq 0.5$. It is clear that $[\tilde{Y}]_\alpha \subseteq [\tilde{Y}]_{1-\alpha}$. In the weak duality theorems we can change the assumptions as follows: $x \in [\tilde{X}]_\alpha$ and $y \in [\tilde{Y}]_\alpha$. Obviously, the statements of the theorems will remain unchanged.

Let us turn to the *strong duality*. We start with the "satisficing" approach to "maximization" or "minimization".

For this purpose, we assume the existence of exogenously given additional fuzzy goals $\tilde{d} \in \mathcal{F}(\mathbb{R})$ and $\tilde{h} \in \mathcal{F}(\mathbb{R})$. The fuzzy goal $\tilde{d}$ is compared to fuzzy values $\tilde{c}_1 x_1 \tilde{+} \cdots \tilde{+} \tilde{c}_n x_n$ of the objective function of the primal FLP problem (P) by fuzzy relation $\gtrsim^{\min}$. On the other hand, the fuzzy goal $\tilde{h}$ is compared to fuzzy values $\tilde{b}_1 y_1 \tilde{+} \cdots \tilde{+} \tilde{b}_m y_m$ of the objective function of the dual FLP problem (D) by fuzzy relation $\lesssim^{\max}$. In this way we treat the fuzzy objectives as constraints

$$\tilde{c}_1 x_1 \tilde{+} \cdots \tilde{+} \tilde{c}_n x_n \gtrsim^{\min} \tilde{d}, \quad \tilde{b}_1 y_1 \tilde{+} \cdots \tilde{+} \tilde{b}_m y_m \lesssim^{\max} \tilde{h}.$$

By $\tilde{X}^*$ we denote the satisficing solution of the primal FLP problem (P), defined by Definition 5.28, by $\tilde{Y}^*$; the satisficing solution of the dual FLP problem (D) is denoted. Clearly, $\tilde{X}^*$ is a fuzzy subset of $\mathbb{R}^n$, $\tilde{Y}^*$ is a fuzzy subset of $\mathbb{R}^m$ and, moreover, $\tilde{X}^* \subseteq \tilde{X}$ and $\tilde{Y}^* \subseteq \tilde{Y}$.

Theorem 5.38. First Strong Duality Theorem. *Let $\tilde{c}_j$, $\tilde{a}_{ij}$ and $\tilde{b}_i$ be fuzzy quantities for all $i \in \mathcal{M}$ and $j \in \mathcal{N}$. Let $\tilde{d}, \tilde{h} \in \mathcal{F}(\mathbb{R})$ be fuzzy goals with the membership functions $\mu_{\tilde{d}}$ and $\mu_{\tilde{h}}$ satisfying the following conditions,*

$$\begin{aligned} &\text{both } \mu_{\tilde{d}} \text{ and } \mu_{\tilde{h}} \text{ are upper semicontinuous,} \\ &\mu_{\tilde{d}} \text{ is strictly increasing, } \mu_{\tilde{h}} \text{ is strictly decreasing,} \qquad (5.72) \\ &\lim_{t \to -\infty} \mu_{\tilde{d}}(t) = \lim_{t \to +\infty} \mu_{\tilde{h}}(t) = 0. \end{aligned}$$

Let $G = T = \min$ and $S = \max$. Let $\tilde{\leq}^{\min}$ be the T-fuzzy extension of the binary relation $\leq$ on $\mathbb{R}$ and $\tilde{\leq}^{\max}$ be the S-fuzzy extension of the relation $\leq$ on $\mathbb{R}$. Let $\tilde{X}^*$ be a satisficing solution of FLP problem (5.65), $\tilde{Y}^*$ be a satisficing solution of FLP problem (5.66) and $\alpha \in (0,1)$.
If a vector $x^* = (x_1^*, \ldots, x_n^*)^T \geq 0$ belongs to $[\tilde{X}^*]_\alpha$, then there exists a vector $y^* = (y_1^*, \ldots, y_m^*)^T \geq 0$ that belongs to $[\tilde{Y}^*]_{1-\alpha}$, and

$$\sum_{j \in \mathcal{N}} \tilde{c}_j^{\mathrm{R}}(\alpha) x_j^* = \sum_{i \in \mathcal{M}} \tilde{b}_i^{\mathrm{R}}(\alpha) y_i^*. \tag{5.73}$$

Proof. Let $x^* = (x_1^*, \ldots, x_n^*)^T \geq 0$, $x^* \in [\tilde{X}^*]_\alpha$. By Proposition 5.35 and Theorem 5.26

$$\sum_{j \in \mathcal{N}} \tilde{c}_j^{\mathrm{R}}(\alpha) x_j^* \geq \tilde{d}^L(\alpha), \tag{5.74}$$

$$\sum_{j \in \mathcal{N}} \tilde{a}_{ij}^{\mathrm{L}}(\alpha) x_j^* \leq \tilde{b}_i^{\mathrm{R}}(\alpha), \quad i \in \mathcal{M}. \tag{5.75}$$

Consider the following LP problem,
(P1)

$$\text{maximize } \sum_{j \in \mathcal{N}} \tilde{c}_j^{\mathrm{R}}(\alpha) x_j$$

$$\text{subject to } \sum_{j \in \mathcal{N}} \tilde{a}_{ij}^{\mathrm{L}}(\alpha) x_j \leq \tilde{b}_i^{\mathrm{R}}(\alpha), \quad i \in \mathcal{M},$$

$$x_j \geq 0, \quad j \in \mathcal{N}.$$

By assumptions (5.72) concerning the fuzzy goal $\tilde{d}$, the system of inequalities (5.74) and (5.75) is satisfied if and only if x^* is the optimal solution of (P1). By the standard strong duality theorem for LP, there exists $y^* \in \mathbb{R}^*$ being an optimal solution of the dual problem
(D1)

$$\text{minimize } \sum_{i \in \mathcal{M}} \tilde{b}_i^{\mathrm{R}}(\alpha) y_i$$

$$\text{subject to } \tilde{c}_j^{\mathrm{R}}(\alpha) \leq \sum_{i \in \mathcal{M}} \tilde{a}_{ij}^{\mathrm{L}}(\alpha) y_i, \quad j \in \mathcal{N},$$

$$y_i \geq 0, \quad i \in \mathcal{M},$$

such that (5.73) holds.
    It remains only to prove that $y^* \in [\tilde{Y}^*]_{1-\alpha}$. This fact follows from assumptions (5.72) concerning the fuzzy goal $\tilde{h}$. □

    Notice that in the crisp case, (5.73) is the standard strong duality result for LP.
    Now we turn to the α-efficient approach to optimization of FLP problems.
    By X_α^* we denote the α-efficient solution of the primal FLP problem (P), defined by Definition 5.33; analogously, by Y_α^* the α-efficient solution of the dual FLP problem (D) is denoted.

Theorem 5.39. Second Strong Duality Theorem. *Let $\tilde{c}_j$, $\tilde{a}_{ij}$ and $\tilde{b}_i$ be fuzzy quantities for all $i \in \mathcal{M}$ and $j \in \mathcal{N}$. Let $\lesssim^{\min}$ and $\gtrsim^{\max}$ be fuzzy extensions of the binary relation $\leq$, and $\alpha \in (0,1)$. If $[\tilde{X}]_\alpha$ and $[\tilde{Y}]_{1-\alpha}$ are nonempty, then there exist x^*, an α-efficient solution of FLP problem (P), and y^*, a $(1-\alpha)$-efficient solution of FLP problem (D) such that*

$$\sum_{j \in \mathcal{N}} \tilde{c}_j^{\mathrm{R}}(\alpha)x_j^* = \sum_{i \in \mathcal{M}} \tilde{b}_i^{\mathrm{R}}(\alpha)y_i^*. \tag{5.76}$$

Proof. Consider the following couple of classical LP problems,

$$\begin{aligned}
\text{maximize } &\sum_{j \in \mathcal{N}} \tilde{c}_j^{\mathrm{R}}(\alpha)x_j \\
\text{subject to } &\sum_{j \in \mathcal{N}} \tilde{a}_{ij}^{\mathrm{L}}(\alpha)x_j \leq \tilde{b}_i^{\mathrm{R}}(\alpha), i \in \mathcal{M}, \\
&x_j \geq 0, j \in \mathcal{N},
\end{aligned} \tag{5.77}$$

and

$$\begin{aligned}
\text{minimize } &\sum_{i \in \mathcal{M}} \tilde{b}_i^{\mathrm{R}}(\alpha)y_i \\
\text{subject to } &\quad\tilde{c}_j^{\mathrm{R}}(\alpha) \quad\leq \sum_{i \in \mathcal{M}} \tilde{a}_{ij}^{\mathrm{L}}(\alpha)y_i, j \in \mathcal{N}, \\
&y_i \geq 0, i \in \mathcal{M}.
\end{aligned} \tag{5.78}$$

Evidently, LP problems (5.77) and (5.78) are dual to each other in the usual sense. Then according to the assumptions both problems (5.77) and (5.78) are feasible and by the standard duality theorem for the couple of dual LP problems, there exist x^*, a feasible solution of problem (5.77), and y^*, a feasible solution of problem (5.78), such that (5.73) holds. By Proposition 5.26 it follows that $x^* \in [\tilde{X}]_\alpha$ and $y^* \in [\tilde{Y}]_{1-\alpha}$ and by Theorem 5.37, x^* is an α-efficient solution of FLP problem (P) and y^* is a $(1-\alpha)$-efficient solution of FLP problem (D). $\qquad\square$

Particularly, in the crisp case, Theorem 5.39 is in fact the strong duality result for standard LP. The question arises as to how the theorems could be modified for more general t-norms and t-conorms.

5.8 Extended addition

Up till now, in Proposition 5.23, formulae (5.33), (5.35) and many others we have used addition of fuzzy values by the t-norm $T_M = \min$. In this section, we investigate addition of fuzzy quantities using a more general t-norm T; particularly we denote

$$\tilde{f} = \tilde{c}_1 x_1 \tilde{+}_T \cdots \tilde{+}_T \tilde{c}_n x_n, \tag{5.79}$$

and

$$\tilde{g}_i = \tilde{a}_{i1}x_1 \widetilde{+}_T \cdots \widetilde{+}_T \tilde{a}_{in}x_n, \tag{5.80}$$

for each $x \in \mathbb{R}^n$, where $\tilde{c}_j, \tilde{a}_{ij} \in \mathcal{F}(\mathbb{R})$, for all $i \in \mathcal{M}$, $j \in \mathcal{N}$. The extended addition $\widetilde{+}_T$ in (5.79) and (5.80) is defined by (5.32) and (5.34), respectively, that is, by using the extension principle. The membership functions of (5.79) and (5.80) are defined as follows.

$$\mu_{\tilde{f}}(t) = \sup\{T(\mu_{\tilde{c}_1}(c_1), \ldots, \mu_{\tilde{c}_n}(c_n)) | t = c_1x_1 + \cdots + c_nx_n\}, \tag{5.81}$$

$$\mu_{\tilde{g}_i}(t) = \sup\{T(\mu_{\tilde{a}_{i1}}(a_{i1}), \ldots, \mu_{\tilde{a}_{in}}(a_{in})) | t = a_{i1}x_1 + \cdots + a_{in}x_n\}. \tag{5.82}$$

Formulae (5.79), (5.80) or (5.81), (5.82) can be difficult to obtain, however, in some special cases analytical formulae can be derived.

For the sake of brevity we deal only with (5.79); formula (5.80) can be obtained analogously. We derive special formulae for a broad class of fuzzy values (i.e., coefficients of the FLP problem) generated by the same functions. Let $\Phi, \Psi : (0, +\infty) \to [0, 1]$ be nonincreasing, semistrictly quasiconcave and upper-semicontinuous functions. Given $\gamma, \delta \in (0, +\infty)$, define functions $\Phi_\gamma, G_\delta : (0, +\infty) \to [0, 1]$ for $x \in (0, +\infty)$ by

$$\Phi_\gamma(x) = \Phi\left(\frac{x}{\gamma}\right), \quad \Psi_\delta(x) = \Psi\left(\frac{x}{\delta}\right).$$

Let $l_j, r_j \in \mathbb{R}$ such that $l_j \le r_j$, let $\gamma_j, \delta_j \in (0, +\infty)$ and let

$$\tilde{c}_j = (l_j, r_j, \Phi_{\gamma_j}, \Psi_{\delta_j}), \quad j \in \mathcal{N},$$

denote fuzzy intervals with the membership functions given by

$$\mu_{\tilde{c}_j}(x) = \begin{cases} \Phi_{\gamma_j}(l_j - x) & \text{if } x \in (-\infty, l_j), \\ 1 & \text{if } x \in [l_j, r_j], \\ \Psi_{\delta_j}(x - r_j) & \text{if } x \in (r_j, +\infty). \end{cases} \tag{5.83}$$

The following proposition shows that $\tilde{c}_1x_1 \widetilde{+}_T \cdots \widetilde{+}_T \tilde{c}_nx_n$ is a closed fuzzy quantity of the same type for particular t-norms T. The proof is straightforward and is omitted here.

Proposition 5.40. *Let $\tilde{c}_j = (l_j, r_j, \Phi_{\gamma_j}, \Psi_{\delta_j})$, $j \in \mathcal{N}$, be fuzzy quantities with the membership functions given by (5.83). For $x = (x_1, \ldots, x_n)^T \in \mathbb{R}^n$, $x_j \ge 0$ for all $j \in \mathcal{N}$, define I_x by*

$$I_x = \{j \mid x_j > 0, j \in \mathcal{N}\}.$$

Then

$$\tilde{c}_1x_1 \widetilde{+}_{T_M} \cdots \widetilde{+}_{T_M} \tilde{c}_nx_n = (l, r, \Phi_{l_M}, \Psi_{r_M}), \tag{5.84}$$

$$\tilde{c}_1x_1 \widetilde{+}_{T_D} \cdots \widetilde{+}_{T_D} \tilde{c}_nx_n = (l, r, \Phi_{l_D}, \Psi_{r_D}), \tag{5.85}$$

where T_M is the minimum t-norm, T_D is the drastic product and

$$l = \sum_{j \in I_x} l_j x_j, \qquad r = \sum_{j \in I_x} r_j x_j,$$

$$l_M = \sum_{j \in I_x} \frac{\gamma_j}{x_j}, \qquad r_M = \sum_{j \in I_x} \frac{\delta_j}{x_j},$$

$$l_D = \max\left\{ \frac{\gamma_j}{x_j} \mid j \in I_x \right\}, \qquad r_D = \max\left\{ \frac{\delta_j}{x_j} \mid j \in I_x \right\}.$$

If all $\tilde{c}_j$ are $(\mathcal{L}, \mathcal{R})$-fuzzy intervals, then an analogous and more specific result can be obtained. Let $l_j, r_j \in \mathbb{R}$ with $l_j \leq r_j$, let $\gamma_j, \delta_j \in [0, +\infty)$ and let $\mathcal{L}, \mathcal{R}$ be nonincreasing, semistrictly quasiconcave, upper-semicontinuous functions from $(0, 1]$ into $[0, +\infty)$, Moreover, assume that $\mathcal{L}(1) = \mathcal{R}(1) = 0$, and define $\mathcal{L}(0) = \lim_{x \to 0} \mathcal{L}(x)$, $\mathcal{R}(0) = \lim_{x \to 0} \mathcal{R}(x)$.

Let

$$\tilde{c}_j = (l_j, r_j, \gamma_j, \delta_j)_{\mathcal{L}\mathcal{R}}$$

be an $(\mathcal{L}, \mathcal{R})$-fuzzy interval given by the membership function defined for each $x \in \mathbb{R}$ and for every $j \in \mathcal{N}$, by

$$\mu_{\tilde{c}_j}(x) = \begin{cases} \mathcal{L}^{(-1)}\left(\frac{l_j - x}{\gamma_j}\right) & \text{if } x \in (l_j - \gamma_j, l_j), \gamma_j > 0, \\ 1 & \text{if } x \in [l_j, r_j], \\ \mathcal{R}^{(-1)}\left(\frac{x - r_j}{\delta_j}\right) & \text{if } x \in (r_j, r_j + \delta_j), \delta_j > 0, \\ 0 & \text{otherwise,} \end{cases} \qquad (5.86)$$

where $\mathcal{L}^{(-1)}, \mathcal{R}^{(-1)}$ are pseudo-inverse functions of $\mathcal{L}, \mathcal{R}$, respectively. We obtain the following result.

Proposition 5.41. Let $\tilde{c}_j = (l_j, r_j, \gamma_j, \delta_j)_{\mathcal{L}\mathcal{R}}$, $j \in \mathcal{N}$, be $(\mathcal{L}, \mathcal{R})$-fuzzy intervals with the membership functions given by (5.86) and let $x = (x_1, \ldots, x_n)^T \in \mathbb{R}^n$, $x_j \geq 0$ for all $j \in \mathcal{N}$. Then

$$\tilde{c}_1 x_1 \tilde{+}_{T_M} \cdots \tilde{+}_{T_M} \tilde{c}_n x_n = (l, r, A_M, B_M)_{\mathcal{L}\mathcal{R}}, \qquad (5.87)$$

$$\tilde{c}_1 x_1 \tilde{+}_{T_D} \cdots \tilde{+}_{T_D} \tilde{c}_n x_n = (l, r, A_D, B_D)_{\mathcal{L}\mathcal{R}}, \qquad (5.88)$$

where T_M is the minimum t-norm, T_D is the drastic product and

$$l = \sum_{j \in \mathcal{N}} l_j x_j, \qquad r = \sum_{j \in \mathcal{N}} r_j x_j,$$

$$A_M = \sum_{j \in \mathcal{N}} \gamma_j x_j, \qquad B_M = \sum_{j \in \mathcal{N}} \delta_j x_j,$$

$$A_D = \max\{\gamma_j \mid j \in \mathcal{N}\}, \qquad B_D = \max\{\delta_j \mid j \in \mathcal{N}\}.$$

The results (5.85) and (5.88) in Propositions 5.40 and 5.41, respectively, can be extended as follows; see also [73].

Proposition 5.42. *Let T be a continuous Archimedean t-norm with an additive generator f. Let $\Phi : (0, +\infty) \rightarrow [0, 1]$ be defined for each $x \in (0, +\infty)$ as*

$$\Phi(x) = f^{(-1)}(x).$$

Let $\tilde{c}_j = (l_j, r_j, \Phi_{\gamma_j}, \Phi_{\delta_j})$, $j \in \mathcal{N}$, be closed fuzzy intervals with the membership functions given by (5.83) and let $x = (x_1, \dots, x_n)^T \in \mathbb{R}^n$, $x_j \geq 0$ for all $j \in \mathcal{N}$, $I_x = \{j \mid x_j > 0, j \in \mathcal{N}\}$. Then

$$\tilde{c}_1 x_1 \tilde{+}_T \cdots \tilde{+}_T \tilde{c}_n x_n = (l, r, \Phi_{l_D}, \Phi_{r_D}),$$

where

$$l = \sum_{j \in I_x} l_j x_j, \quad r = \sum_{j \in I_x} r_j x_j,$$

$$l_D = \max \left\{ \frac{\gamma_j}{x_j} \mid j \in I_x \right\}, \quad r_D = \max \left\{ \frac{\delta_j}{x_j} \mid j \in I_x \right\}.$$

For a continuous Archimedean t-norm T and closed fuzzy interval $\tilde{c}_j$ satisfying the assumptions of Proposition 5.42, we obtain easily

$$\tilde{c}_1 x_1 \tilde{+}_T \cdots \tilde{+}_T \tilde{c}_n x_n = \tilde{c}_1 x_1 \tilde{+}_{T_D} \cdots \tilde{+}_{T_D} \tilde{c}_n x_n,$$

which means that we obtain the same fuzzy linear function based on an arbitrary t-norm T' such that $T' \leq T$.

The result that follows generalizes a result concerning the addition of closed fuzzy intervals based on continuous Archimedean t-norms.

Proposition 5.43. *Let T be a continuous Archimedean t-norm with an additive generator f. Let $K : [0, +\infty) \rightarrow [0, +\infty)$ be a continuous convex function with $K(0) = 0$. Let $\alpha \in (0, +\infty)$ and*

$$\Phi_\alpha(x) = f^{(-1)} \left(\alpha K \left(\frac{x}{\alpha} \right) \right)$$

for all $x \in [0, +\infty)$. Let $\tilde{c}_j = (l_j, r_j, \Phi_{\gamma_j}, \Phi_{\delta_j})$, $j \in \mathcal{N}$, be closed fuzzy intervals with the membership functions given by (5.83) and let $x = (x_1, \dots, x_n)^T \in \mathbb{R}^n$, $x_j \geq 0$ for all $j \in \mathcal{N}$, $I_x = \{j \mid x_j > 0, j \in \mathcal{N}\}$. Then

$$\tilde{c}_1 x_1 \tilde{+}_T \cdots \tilde{+}_T \tilde{c}_n x_n = (l, r, \Phi_{l_K}, \Phi_{r_K}),$$

where

$$l = \sum_{j \in I_x} l_j x_j, \quad r = \sum_{j \in I_x} r_j x_j, \tag{5.89}$$

$$l_K = \sum_{j \in I_x} \frac{\gamma_j}{x_j}, \quad r_K = \sum_{j \in I_x} \frac{\delta_j}{x_j}.$$

A simple consequence can be obtained from Proposition 5.42.

The sum based on the product t-norm T_P of Gaussian fuzzy numbers is again a Gaussian fuzzy number. Indeed, the additive generator f of the product t-norm T_P is given by $f(x) = -\log(x)$. Let $K(x) = x^2$. Then

$$\Phi_\alpha(x) = f^{(-1)}\left(\alpha K\left(\frac{x}{\alpha}\right)\right) = e^{-\frac{x^2}{\alpha}}.$$

By Proposition 5.43 we obtain the formula.

A similar approach based on centered fuzzy numbers is mentioned later in this chapter; see also [76], [78] and [135].

5.9 Special models of FLP

Three types of the FLP problem known from the literature are investigated in this section. We start with the oldest version of FLP problems, originally called the fuzzy (linear) programming problem; see [197]. Later on (e.g., see [81]), this problem was named the flexible LP problem.

5.9.1 Flexible linear programming

Flexible linear programming refers to the approach to LP problems allowing for a kind of flexibility of the objective function and constraints in the standard LP problem (5.29). Consider

$$\begin{aligned}
\text{maximize } & c_1 x_1 + \cdots + c_n x_n \\
\text{subject to } & a_{i1} x_1 + \cdots + a_{in} x_n \leq b_i, \quad i \in \mathcal{M}, \\
& x_j \geq 0, \quad j \in \mathcal{N}.
\end{aligned} \tag{5.90}$$

The values of parameters c_j, a_{ij} and b_i in (5.90) are supposed to be subjected to some uncertainty. By nonnegative values p_i, $i \in \{0\} \cup \mathcal{M}$, admissible violations of the objective and constraints are (subjectively) chosen and introduced to the original model (5.90).

An aspiration level $d_0 \in \mathbb{R}$ is (subjectively) determined such that the decision maker (DM) is fully satisfied on the condition that the value of the objective function is greater than or equal to d_0. On the other hand, if the objective function attains a value smaller than $d_0 - p_0$, then DM is fully dissatisfied. Within the interval $(d_0 - p_0, d_0)$, the satisfaction of DM increases (e.g., linearly) from 0 to 1. Under these assumptions a membership function $\mu_{\tilde{d}}$ of the fuzzy goal $\tilde{d}$ could be defined as follows,

$$\mu_{\tilde{d}}(t) = \begin{cases} 1 & \text{if } t \geq d_0, \\ 1 + \frac{t - d_0}{p_0} & \text{if } d_0 - p_0 \leq t < d_0, \\ 0 & \text{otherwise.} \end{cases} \tag{5.91}$$

Now, let for the ith constraint function of (5.90), $i \in \mathcal{M}$; a right-hand side $b_i \in \mathbb{R}$ is known such that then the decision maker is fully satisfied on condition that the left-hand side is less than or equal to this value. On the other hand, if the objective function is greater than $b_i + p_i$, then the DM is fully dissatisfied. Within the interval $(b_i, b_i + p_i)$, the satisfaction of DM decreases (linearly) from 1 to 0. Under these assumptions the membership function $\mu_{\tilde{b}_i}$ of the fuzzy right-hand side $\tilde{b}_i$ is defined as

$$\mu_{\tilde{b}_i}(t) = \begin{cases} 1 & \text{if } t \leq b_i, \\ 1 - \frac{t-b_i}{p_i} & \text{if } b_i \leq t < b_i + p_i, \\ 0 & \text{otherwise.} \end{cases} \tag{5.92}$$

The relationship between the objective function and constraints in the flexible LP problem is symmetric; i.e., there is no a difference between the former and the latter. "Maximization" is understood as finding a vector $x \in \mathbb{R}^n$ such that the membership grade of the intersection of fuzzy sets (5.91) and (5.92) is maximized. This problem is equivalent to the following optimization problem,

$$\text{maximize } \lambda$$
$$\text{subject to } \mu_{\tilde{d}}\left(\sum_{j \in \mathcal{N}} c_j x_j\right) \geq \lambda,$$
$$\mu_{\tilde{b}_i}\left(\sum_{j \in \mathcal{N}} a_{ij} x_j\right) \geq \lambda, \quad i \in \mathcal{M}, \tag{5.93}$$
$$0 \leq \lambda \leq 1,$$
$$x_j \geq 0, \quad j \in \mathcal{N}.$$

Problem (5.93) can easily be transformed to the equivalent LP problem:

$$\text{maximize } \lambda$$
$$\text{subject to } \sum_{j \in \mathcal{N}} c_j x_j \geq d_0 + \lambda p_0,$$
$$\sum_{j \in \mathcal{N}} a_{ij} x_j \leq b_i + (1 - \lambda)p_i, \quad i \in \mathcal{M}, \tag{5.94}$$
$$0 \leq \lambda \leq 1,$$
$$x_j \geq 0, \quad j \in \mathcal{N}.$$

Now, consider a more specific FLP problem:

$$\text{maximize } c_1 x_1 + \cdots + c_n x_n$$
$$\text{subject to } a_{i1} x_1 + \cdots + a_{in} x_n \tilde{\leq}^T \tilde{b}_i, \quad i \in \mathcal{M}, \tag{5.95}$$
$$x_j \geq 0, \quad j \in \mathcal{N},$$

where c_j, a_{ij} and b_i are crisp numbers, whereas $\tilde{d}$ and $\tilde{b}_i$ are fuzzy quantities defined by (5.91) and (5.92). Moreover, $\tilde{\leq}^T$ is a T-fuzzy extension of the usual inequality relation $\leq$, with $T = \min$. It turns out that the vector $x \in \mathbb{R}^n$ is an optimal solution of the flexible LP problem (5.94) if and only if it is a max-satisficing solution of FLP problem (5.95). This result follows directly from Proposition 5.32.

5.9.2 Interval linear programming

In this subsection we apply the results of this chapter to a special case of the FLP problem, the interval linear programming problem investigated already in Chapters 2, 3 and 4. By *interval linear programming problem (ILP)* we understand the following FLP problem,

$$
\begin{aligned}
\text{maximize } & \tilde{c}_1 x_1 \tilde{+} \cdots \tilde{+} \tilde{c}_n x_n \\
\text{subject to } & \tilde{a}_{i1} x_1 \tilde{+} \cdots \tilde{+} \tilde{a}_{in} x_n \tilde{R} \quad \tilde{b}_i, \quad i \in \mathcal{M}, \\
& x_j \geq 0, \quad j \in \mathcal{N},
\end{aligned}
\tag{5.96}
$$

where $\tilde{c}_j$, $\tilde{a}_{ij}$ and $\tilde{b}_i$ are considered to be compact intervals in $\mathbb{R}$. I.e., $\tilde{c}_j = [\underline{c}_j, \overline{c}_j]$, $\tilde{a}_{ij} = [\underline{a}_{ij}, \overline{a}_{ij}]$ and $\tilde{b}_i = [\underline{b}_i, \overline{b}_i]$, where $\underline{c}_j$, $\overline{c}_j$, $\underline{a}_{ij}$, $\overline{a}_{ij}$ and $\underline{b}_i$, $\overline{b}_i$ are lower and upper bounds of the corresponding intervals, respectively (see also Chapter 2). Let the membership functions of $\tilde{c}_j$, $\tilde{a}_{ij}$ and $\tilde{b}_i$ be the characteristic functions of the intervals. I.e., $\chi_{[\underline{c}_j, \overline{c}_j]} : \mathbb{R} \rightarrow [0,1]$, $\chi_{[\underline{a}_{ij}, \overline{a}_{ij}]} : \mathbb{R} \rightarrow [0,1]$ and $\chi_{[\underline{b}_i, \overline{b}_i]} : \mathbb{R} \rightarrow [0,1]$, $i \in \mathcal{M}$, $j \in \mathcal{N}$.

Now, we assume that R is the usual binary relation $\leq$, and $A = T = \min$, $S = \max$. The fuzzy relation $\tilde{R}$ is the fuzzy extension of a valued relation $\leq$. We consider six fuzzy relations $\tilde{R}$, extensions of the binary relation $\leq$, defined by (5.19), (5.20) and by (5.22)–(5.25); i.e.,

$$
\tilde{R} \in \left\{ \tilde{\leq}^{\min}, \tilde{\leq}^{\max}, \tilde{\leq}^{T,S}, \tilde{\leq}_{T,S}, \tilde{\leq}^{S,T}, \tilde{\leq}_{S,T} \right\}.
$$

Then by Proposition 5.26 we obtain six types of feasible solutions of ILP problem (5.96).

(i)

$$
X_{\tilde{\leq}^{\min}} = \left\{ x \in \mathbb{R}^n \Big| \sum_{j=1}^{n} \underline{a}_{ij} x_j \leq \overline{b}_i, \ x_j \geq 0, j \in \mathcal{N} \right\}.
\tag{5.97}
$$

(ii)

$$
X_{\tilde{\leq}^{\max}} = \left\{ x \in \mathbb{R}^n \Big| \sum_{j=1}^{n} \overline{a}_{ij} x_j \leq \underline{b}_i, \ x_j \geq 0, j \in \mathcal{N} \right\}.
\tag{5.98}
$$

(iii)

$$
X_{\tilde{\leq}^{T,S}} = X_{\tilde{\leq}_{T,S}} = \left\{ x \in \mathbb{R}^n \Big| \sum_{j=1}^{n} \overline{a}_{ij} x_j \leq \overline{b}_i, \ x_j \geq 0, j \in \mathcal{N} \right\}.
\tag{5.99}
$$

(iv)

$$X_{\tilde{\leq}s.T} = X_{\tilde{\leq}_{s.T}} = \left\{ x \in \mathbb{R}^n \mid \sum_{j=1}^n \underline{a}_{ij} x_j \leq \underline{b}_i, \ x_j \geq 0, \ j \in \mathcal{N} \right\}. \quad (5.100)$$

Clearly, feasible solutions (5.97)–(5.100) are crisp subsets of $\mathbb{R}^n$; moreover, they all are polyhedral.

In order to find, e.g., a satisficing solution of ILP problem (5.96), we consider a fuzzy goal $\tilde{d} \in \mathcal{F}(\mathbb{R})$ and $\tilde{R}_0$, a fuzzy extension of the usual binary relation $\geq$ for comparing the objective with the fuzzy goal.

In the following proposition we show that if the feasible solution of the ILP problem is crisp then its max-satisficing solution is the same as the set of all classical optimal solutions of the LP problem of maximizing a particular crisp objective over the set of feasible solutions.

Proposition 5.44. *Let X be a crisp feasible solution of ILP problem (5.96). Let $\tilde{d} \in \mathcal{F}(\mathbb{R})$ be a fuzzy goal with the membership function $\mu_{\tilde{d}}$ satisfying conditions (5.53). Let $G_A = G = T = \min$ and $S = \max$.*

(i) If $\tilde{R}_0$ is $\geq^{\min}$, then the set of all max-satisficing solutions of ILP problem (5.96) coincides with the set of all optimal solution of the problem

$$\begin{aligned} maximize \ \textstyle\sum_{j=1}^n \overline{c}_j x_j \\ subject \ to \ x \in X. \end{aligned} \quad (5.101)$$

(ii) If $\tilde{R}_0$ is $\geq^{\max}$, then the set of all max-satisficing solutions of ILP problem (5.96) coincides with the set of all optimal solution of the problem

$$\begin{aligned} maximize \ \textstyle\sum_{j=1}^n \underline{c}_j x_j \\ subject \ to \ x \in X. \end{aligned}$$

Proof. (i) Let $x \in X$ be a max-satisficing solution of ILP problem (5.96), $\underline{c} = \sum_{j=1}^n \underline{c}_j x_j$, $\overline{c} = \sum_{j=1}^n \overline{c}_j x_j$. Our assumptions give

$$\begin{aligned} \mu_{\tilde{\geq}T}(\tilde{c}_1 x_1 \tilde{+} \cdots \tilde{+} \tilde{c}_n x_n, \tilde{d}) \\ = \sup\{\min\{\mu_{\tilde{c}_1 x_1 \tilde{+} \cdots \tilde{+} \tilde{c}_n x_n}(u), \mu_{\tilde{d}}(v)\} \mid u \geq v\} \\ = \sup\{\min\{\chi_{[\underline{c},\overline{c}]}(u), \mu_{\tilde{d}}(v)\} \mid u \geq v\} \\ = \mu_{\tilde{d}}\left(\textstyle\sum_{j=1}^n \overline{c}_j x_j\right). \end{aligned}$$

Hence, x is an optimal solution of (5.101). Conversely, if $x \in X$ is an optimal solution of (5.101), then by Definition 5.28 and by (5.53), x is a max-satisficing solution of problem (5.96).

(ii) Analogously to the proof of (i), we have

$$\mu_{\geq_s}(\tilde{c}_1 x_1 \dotplus \cdots \dotplus \tilde{c}_n x_n, \tilde{d})$$

$$= \inf\{\max\{1 - \mu_{\tilde{c}_1 x_1 \dotplus \cdots \dotplus \tilde{c}_n x_n}(u), 1 - \mu_{\tilde{d}}(v)\} \mid u \geq v\}$$

$$= \inf\{1 - \min\{\chi_{[\underline{c}, \overline{c}]}(u), \mu_{\tilde{d}}(v)\} \mid u \leq v\}$$

$$= \mu_{\tilde{d}}\left(\sum_{j=1}^n \underline{c}_j x_j\right).$$

By the same arguments as in (i) we conclude the proof.   □

We close this section with several observations concerning duality of ILP problems.

Let the primal ILP problem (P) be problem (5.96) with $\tilde{R}$ be $\leq^{\min}$; i.e., (5.65) holds. Then the dual ILP problem (D) is (5.66). Clearly, the feasible solution $X_{\leq \min}$ of (P) is defined by (5.97) and the feasible solution $Y_{\geq \max}$ of the dual problem (D) can be derived from (5.98) as

$$Y_{\geq \max} = \left\{ y \in \mathbb{R}^m \mid \sum_{i=1}^m \underline{a}_{ij} y_i \geq \overline{c}_j, \, y_i \geq 0, \, i \in \mathcal{M} \right\}.$$

Notice that the problems

$$\text{maximize } \sum_{j=1}^n \overline{c}_j x_j$$
$$\text{subject to } x \in X_{\leq \min}$$

and

$$\text{minimize } \sum_{i=1}^m \overline{b}_i y_i$$
$$\text{subject to } y \in Y_{\geq \max}$$

are dual to each other in the usual sense if and only if $\underline{c}_j = \overline{c}_j$ and $\underline{b}_i = \overline{b}_i$ for all $i \in \mathcal{M}$ and $j \in \mathcal{N}$.

5.9.3 FLP problems with centered parameters

An interesting class of FLP problems can be obtained if the parameters of the FLP problem are fuzzy sets called $\mathcal{B}$-fuzzy intervals; see [77], [135].

Definition 5.45. *A fuzzy set A given by the membership function $\mu_A : \mathbb{R} \to [0, 1]$ is called a* generator *in $\mathbb{R}$ if*

(i) $0 \in \text{Core}(A)$,
(ii) μ_A is quasiconcave on $\mathbb{R}$.

Notice that each generator is a special fuzzy interval A that satisfies (i).

Definition 5.46. *A set $\mathcal{B}$ of generators in $\mathbb{R}$ is called a* basis of generators *in $\mathbb{R}$ if*

(i) $\chi_{\{0\}} \in \mathcal{B}$, $\chi_{\mathbb{R}} \in \mathcal{B}$,
(ii) If $f, g \in \mathcal{B}$ then $\max\{f, g\} \in \mathcal{B}$ and $\min\{f, g\} \in \mathcal{B}$.

Definition 5.47. *Let $\mathcal{B}$ be a basis of generators. A fuzzy set A given by the membership function $\mu_A : \mathbb{R} \to [0, 1]$ is called a $\mathcal{B}$-fuzzy interval if there exist $a_A \in \mathbb{R}$ and $g_A \in \mathcal{B}$ such that for each $x \in \mathbb{R}$,*

$$\mu_A(x) = g_A(x - a_A).$$

The set of all $\mathcal{B}$-fuzzy intervals is denoted by $\mathcal{F}_{\mathcal{B}}(\mathbb{R})$. Each $A \in \mathcal{F}_{\mathcal{B}}(\mathbb{R})$ is represented by a pair (a_A, g_A); we write $A = (a_A, g_A)$. An ordering relation $\leq_{\mathcal{B}}$ is defined on $\mathcal{F}_{\mathcal{B}}(\mathbb{R})$ as follows. For $A, B \in \mathcal{F}_{\mathcal{B}}(\mathbb{R})$, $A = (a_A, g_A)$ and $B = (a_B, g_B)$, we write $A \leq_{\mathcal{B}} B$ if and only if

$$(a_A < a_B) \text{ or } (a_A = a_B \text{ and } g_A \leq g_B). \tag{5.102}$$

Notice that $\leq_{\mathcal{B}}$ is a partial ordering on $\mathcal{F}_{\mathcal{B}}(\mathbb{R})$. The following proposition is a simple consequence of Definition 5.46.

Proposition 5.48. *A pair $(\mathcal{B}, \leq)$, where $\mathcal{B}$ is a basis of generators and $\leq$ is the pointwise ordering of functions, is a lattice with the maximal element $\chi_{\mathbb{R}}$ and minimal element $\chi_{\{0\}}$.*

Example 5.49. The following sets of functions form a basis of generators in $\mathbb{R}$.

(i) $\mathcal{B}_D = \{\chi_{\{0\}}, \chi_{\mathbb{R}}\}$—discrete basis,
(ii) $\mathcal{B}_I = \{\chi_{[a,b]} \mid -\infty \leq a \leq 0 \leq b \leq +\infty\}$—interval basis,
(iii) $\mathcal{B}_G = \{\mu_d \mid \mu_d(x) = g^{(-1)}(|x|/d), \ x \in \mathbb{R}, \ d > 0\} \cup \{\chi_{\{0\}}, \chi_{\mathbb{R}}\}$, where $g : (0, 1] \to [0, +\infty)$ is a nonincreasing nonconstant function, $g(1) = 0$ and $g(0) = \lim_{x \to 0} g(x)$. Obviously, the relation $\leq$ between function values is a linear ordering on $\mathcal{B}_G$.

Proposition 5.50. *Let $\mathcal{F}_{\mathcal{B}_G}(\mathbb{R})$ be the set of all $\mathcal{B}_G$-fuzzy intervals, where $\mathcal{B}_G$ is the basis from Example 5.49. Then the relation $\leq_{\mathcal{B}_G}$ is a linear ordering on $\mathcal{F}_{\mathcal{B}_G}(\mathbb{R})$.*

We can extend this result as follows. Let $\mathcal{B}$ be a basis of generators and $\leq_{\mathcal{B}}$ be a partial ordering on the set $\mathcal{F}_{\mathcal{B}}(\mathbb{R})$ defined by (5.102) in Definition 5.47. If $\mathcal{B}$ is linearly ordered by $\subseteq$, then $\mathcal{F}_{\mathcal{B}}(\mathbb{R})$ is linearly ordered by $\leq_{\mathcal{B}}$. It follows that each $\tilde{c} \in \mathcal{F}_{\mathcal{B}}(\mathbb{R})$ can be uniquely represented by a pair (c, μ), where $c \in \mathbb{R}$ and $\mu \in \mathcal{B}$ such that

$$\mu_{\tilde{c}}(t) = \mu(c - t),$$

therefore we can write $\tilde{c} = (c, \mu)$.

Let $\circ$ be either addition or multiplication, arithmetic operations on $\mathbb{R}$, and $\star$ be either min or max operations on $\mathcal{B}$. On $\mathcal{F}_{\mathcal{B}}(\mathbb{R})$ we introduce the following operations,

$$(a, f) \circ^{(\star)} (b, g) = (a \circ b, f \star g) \tag{5.103}$$

for all $(a, f), (b, g) \in \mathcal{F}_\mathcal{B}(\mathbb{R})$.

Evidently, the pairs of operations $(+^{(\min)}, \cdot^{(\min)})$, $(+^{(\min)}, \cdot^{(\max)})$, $(+^{(\max)}, \cdot^{(\min)})$, and $(+^{(\max)}, \cdot^{(\max)})$, are distributive. For more properties, see [78].

Now, consider $\mathcal{B}$-fuzzy intervals: $\tilde{c}_j = (c_j, f_j)$, $\tilde{a}_{ij} = (a_{ij}, g_{ij})$, $\tilde{b}_i = (b_i, h_i)$, $\tilde{c}_j, \tilde{a}_{ij}, \tilde{b}_i \in \mathcal{F}_\mathcal{B}(\mathbb{R})$, $i \in \mathcal{M}$, $j \in \mathcal{N}$. Let $\diamond$ and $\star$ be either min or max operations on $\mathcal{B}$. Consider the following optimization problem,

$$\text{maximize } \tilde{c}_1 \cdot^{(\diamond)} \tilde{x}_1 +^{(\star)} \cdots +^{(\star)} \tilde{c}_n \cdot^{(\diamond)} \tilde{x}_n$$
$$\text{subject to } \tilde{a}_{i1} \cdot^{(\diamond)} \tilde{x}_1 +^{(\star)} \cdots +^{(\star)} \tilde{a}_{in} \cdot^{(\diamond)} \tilde{x}_n \leq_\mathcal{B} \tilde{b}_i, \quad i \in \mathcal{M}, \quad (5.104)$$
$$\tilde{x}_j \geq_\mathcal{B} \tilde{0}, \quad j \in \mathcal{N}.$$

In (5.104), maximization is performed with respect to the ordering $\leq_\mathcal{B}$; moreover, $\tilde{x}_j = (x_j, \xi_j)$, where $x_j \in \mathbb{R}$ and $\xi_j \in \mathcal{B}$, $\tilde{0} = (0, \chi_{\{0\}})$. The inequalities $\tilde{x}_j \geq_\mathcal{B} \tilde{0}$, $j \in \mathcal{N}$, are equivalent to $x_j \geq 0$, $j \in \mathcal{N}$. Now, we define feasible and optimal solutions.

A *feasible solution* of the problem (5.104) is a vector

$$(\tilde{x}_1, \tilde{x}_2, \ldots, \tilde{x}_n) \in \mathcal{F}_\mathcal{B}(\mathbb{R}) \times \mathcal{F}_\mathcal{B}(\mathbb{R}) \times \cdots \times \mathcal{F}_\mathcal{B}(\mathbb{R}),$$

satisfying the constraints

$$\tilde{a}_{i1} \cdot^{(\diamond)} \tilde{x}_1 +^{(\star)} \cdots +^{(\star)} \tilde{a}_{in} \cdot^{(\diamond)} \tilde{x}_n \leq_\mathcal{B} \tilde{b}_i, \quad i \in \mathcal{M},$$
$$\tilde{x}_j \geq_\mathcal{B} \tilde{0}, \quad j \in \mathcal{N}.$$

The set of all feasible solutions of (5.104) is denoted by $X_\mathcal{B}$.

An *optimal solution* of the problem (5.104) is a vector

$$(\tilde{x}_1^*, \tilde{x}_2^*, \ldots, \tilde{x}_n^*)^T \in \mathcal{F}_\mathcal{B}(\mathbb{R}) \times \mathcal{F}_\mathcal{B}(\mathbb{R}) \times \cdots \times \mathcal{F}_\mathcal{B}(\mathbb{R})$$

such that

$$\tilde{z}^* = \tilde{c}_1 \cdot^{(\diamond)} \tilde{x}_1^* +^{(\star)} \cdots +^{(\star)} \tilde{c}_n \cdot^{(\diamond)} \tilde{x}_n^*$$

is the maximal element (with respect to the ordering $\leq_\mathcal{B}$) of the set

$$X_\mathcal{B}^* = \{\tilde{z} \mid \tilde{z} = \tilde{c}_1 \cdot^{(\diamond)} \tilde{x}_1 +^{(\star)} \cdots +^{(\star)} \tilde{c}_n \cdot^{(\diamond)} \tilde{x}_n, (\tilde{x}_1, \tilde{x}_2, \ldots, \tilde{x}_n)^T \in X_\mathcal{B}\}.$$

For each of four combinations of min and max in the operations $\cdot^{(\diamond)}$ and $+^{(\star)}$, (5.104) is a particular optimization problem. We can easily derive the following result.

Proposition 5.51. *Let $\mathcal{B}$ be a linearly ordered basis of generators. Let $(\tilde{x}_1^*, \tilde{x}_2^*, \ldots, \tilde{x}_n^*)^T \in \mathcal{F}_\mathcal{B}(\mathbb{R})^n$ be an optimal solution of (5.104), where $\tilde{x}_j^* = (x_j^*, \xi_j^*)$, $j \in \mathcal{N}$. Then the vector $x^* = (x_1^*, \ldots, x_n^*)$ is an optimal solution of the following LP problem,*

$$\text{maximize } c_1 x_1 + \cdots + c_n x_n$$
$$\text{subject to } a_{i1} x_1 + \cdots + a_{in} x_n \leq b_i, \quad i \in \mathcal{M}, \quad (5.105)$$
$$x_j \geq 0, \quad j \in \mathcal{N}.$$

Now, by A_x we denote the set of indices of all active constraints of (5.105) at $x = (x_1, \ldots, x_n)$; i.e.,

$$A_x = \{i \in \mathcal{M} \mid a_{i1}x_1 + \cdots + a_{in}x_n = b_i\}.$$

The following proposition gives a necessary condition for the existence of a feasible solution of (5.104). The proof can be found in [77].

Proposition 5.52. *Let $\mathcal{B}$ be a linearly ordered basis of generators. Let $(\tilde{x}_1, \tilde{x}_2, \ldots, \tilde{x}_n)^T \in \mathcal{F}_{\mathcal{B}}(\mathbb{R})^n$ be a feasible solution of (5.104), where $\tilde{x}_j = (x_j, \xi_j)$, $j \in \mathcal{N}$. Then the vector $x = (x_1, \ldots, x_n)^T$ is the feasible solution of the LP problem (5.105) and it holds that*

(i) If $\diamond = \max$ and $\star = \min$, then

$$\min\{a_{ij} \mid j \in \mathcal{N}\} \leq_{\mathcal{B}} b_i \quad \text{for all } i \in A_x;$$

(ii) If $\diamond = \max$ and $\star = \max$, then

$$\max\{a_{ij} \mid j \in \mathcal{N}\} \leq_{\mathcal{B}} b_i \quad \text{for all } i \in A_x.$$

Notice that in this subsection we have presented an alternative approach to FLP problems. Compared to the approach presented before, the decision variables x_j considered here have not been nonnegative numbers. They have been considered as fuzzy intervals of the same type as the corresponding coefficients of the FLP problem. From the computational point of view this approach is simple as it requires solving only a classical LP problem.

5.10 Fuzzy multicriteria linear programming problem

Up till now we have investigated fuzzy linear programming problems with one criterion; the objective function. Our approach can, however, be easily extended to the multicriteria case. The *fuzzy multicriteria linear programming problem (FMLP problem)* associated with FLP problem (5.31) is denoted by:

$$\begin{aligned}
&\text{"maximize" } \tilde{c}_{k1}x_1 \tilde{+} \cdots \tilde{+} \tilde{c}_{kn}x_n, \quad k \in \mathcal{K}, \\
&\text{subject to } (\tilde{a}_{i1}x_1 \tilde{+} \cdots \tilde{+} \tilde{a}_{in}x_n)\tilde{R}_i\tilde{b}_i, \quad i \in \mathcal{M}, \qquad (5.106)\\
&\qquad\qquad x_j \geq 0, \quad j \in \mathcal{N}.
\end{aligned}$$

where $\mathcal{K} = \{1, 2, \ldots, q\}$ is a set of fuzzy criteria and $\mu_{\tilde{c}_{kj}} : \mathbb{R} \to [0,1]$, $\mu_{\tilde{a}_{ij}} : \mathbb{R} \to [0,1]$ and $\mu_{\tilde{b}_i} : \mathbf{R} \to [0,1]$, $k \in \mathcal{K}$, $i \in \mathcal{M}$, $j \in \mathcal{N}$, are membership functions of the fuzzy parameters $\tilde{c}_{kj}$, $\tilde{a}_{ij}$ and $\tilde{b}_i$, respectively.

In order to "maximize" the objective functions we can use similar concepts of "optimal solution" as we have done in the case of one criterion, namely: (1) the satisficing solution, (2) the α-efficient solution. Here, to define a compromise solution of a FMLP problem, we use only the satisficing solution; the other concept can be used analogously.

Hence, for each criterion

$$\tilde{f}_k(x, \tilde{c}_k) = \tilde{c}_{k1}x_1 \tilde{+} \cdots \tilde{+} \tilde{c}_{kn}x_n, \ k \in \mathcal{K}, \tag{5.107}$$

we assume the existence of a given additional goal $\tilde{d}_k \in \mathcal{F}(\mathbb{R})$, the fuzzy set of the real line. The determination of the goal can be essential for the quality of the "optimal" solution. Its meaning, however, depends on the nature of the criteria being in some sense ideal values of the corresponding criteria. The fuzzy value $\tilde{f}_k(x, \tilde{c}_k)$ of the criteria function is compared to the goal $\tilde{d}_k$ by means of a fuzzy relation $\tilde{S}_k$, also given exogenously. Then the fuzzy criteria are treated again as constraints $\tilde{f}_j(x, \tilde{c}_k) \ \tilde{S}_k \ \tilde{d}_k$.

Definition 5.53. *Let f_k, $k \in \mathcal{K}$, be linear functions (5.107). Let $\mu_{\tilde{c}_{kj}} : \mathbb{R} \to [0,1]$ be membership functions of fuzzy parameters $\tilde{c}_{kj}$ and let $\tilde{d}_k \in \mathcal{F}(\mathbb{R})$ be fuzzy subsets of $\mathbb{R}$ called the fuzzy goals, $k \in \mathcal{K}$, $j \in \mathcal{N}$. Furthermore, let $\tilde{S}_k$, $k \in \mathcal{K}$, be fuzzy relations given by the membership functions $\mu_{\tilde{S}_k} : \mathcal{F}(\mathbb{R}) \times \mathcal{F}(\mathbb{R}) \to [0,1]$, and let G_F be an aggregation operator. A fuzzy subset $\tilde{F}$ of $\mathbb{R}^n$ given by the membership function $\mu_{\tilde{F}}$, for all $x \in \mathbb{R}^n$ defined as*

$$\mu_{\tilde{F}}(x) = G_F \left(\mu_{\tilde{S}_1} \left(\tilde{f}_1(x, \tilde{c}_1), \tilde{d}_1 \right), \ldots, \mu_{\tilde{S}_k} \left(\tilde{f}_k(x, \tilde{c}_k), \tilde{d}_k \right) \right) \tag{5.108}$$

is called the criteria fuzzy set *of the FMLP problem (5.106).*

For $k \in \mathcal{K}$, we use the following notation: by $\tilde{F}_k$ we denote a fuzzy set given by the membership function $\mu_{\tilde{F}_k}$, which is defined for all $x \in \mathbb{R}^n$ as

$$\mu_{\tilde{F}_k}(x) = \mu_{\tilde{S}_k} \left(\tilde{f}_k(x, \tilde{c}_k), \tilde{d}_k \right). \tag{5.109}$$

Notice that in the case of the single-criterion problem (i.e., $q = 1$), the aggregation operator G_F is the identity operator and the membership function of the criteria fuzzy set is $\mu_{\tilde{F}}(x) = \mu_{\tilde{S}_1} \left(\tilde{f}_1(x, \tilde{c}_1), \tilde{d}_1 \right)$ for all $x \in \mathbb{R}^n$.

Definition 5.54. *Let f_k, $k \in \mathcal{K}$, g_i, $i \in \mathcal{M}$, be functions, $\tilde{c}_{kj}$, $\tilde{a}_{ij}$ and $\tilde{b}_i$ be fuzzy parameters, and $\tilde{d}_k \in \mathcal{F}(\mathbb{R})$ be the fuzzy goals; furthermore, let $\tilde{R}_i$, $\tilde{S}_k$ be fuzzy relations given by Definitions 5.25 and 5.53. Finally, let G_X, G_F and G be aggregation operators. A fuzzy set $\tilde{X}^*$ given by the membership function $\mu_{\tilde{X}}^*$ for all $x \in \mathbb{R}^n$ as*

$$\mu_{\tilde{X}}^*(x) = G \left(\mu_{\tilde{X}}(x), \mu_{\tilde{F}}(x) \right), \tag{5.110}$$

is called the compromise solution *of the FMLP problem (5.106), where $\mu_{\tilde{F}}$ and $\mu_{\tilde{X}}$ are the membership functions of the criteria fuzzy set and membership function of the feasible solution, respectively.*

For $\alpha \in (0, 1]$ a vector $x \in [\tilde{X}^*]_\alpha$ is called the α-compromise solution of the FMLP problem (5.106).
A vector $x^* \in \mathbb{R}^n$ with the property $\mu_{\tilde{X}}^*(x^*) = Hgt(\tilde{X}^*)$ is called the max-compromise solution.

Notice that the compromise solution $\tilde{X}^*$ of a FMLP problem is a fuzzy subset of $\mathbb{R}^n$. In the case of a single criterion the compromise solution is in fact the satisficing solution by Definition 5.28. Moreover, we obtain $\tilde{X}^* \subseteq \tilde{X}$, where $\tilde{X}$ is a feasible solution.

On the other hand, the α-compromise solution is a vector, as well as the max-compromise solution, which is, in fact, the α-compromise solution with $\alpha = Hgt(\tilde{X}^*)$.

In Definition 5.54, we consider three aggregation operators G_X, G_F and G. The first operator G_X is used for aggregating the individual constraints into the feasible solution by Definition 5.25, the second one, G_F, is used for aggregating the individual criteria and by the third one, G, the criteria and constraints are aggregated. It may be difficult or impossible to choose a proper aggregation operator for combining the criteria and constraints together. In this case we can apply the well-known concept of Pareto-optimal solution adapted to the "new criteria" represented by the membership functions $\mu_{\tilde{X}}$ and $\mu_{\tilde{F}}$; see also [135].

Definition 5.55. *Let $\mu_{\tilde{F}}$ and $\mu_{\tilde{X}}$ be the membership functions of the criteria fuzzy set by Definition 5.53 and the membership function of the feasible solution defined by Definition 5.54, respectively.*
A vector $x^p \in \mathbb{R}^n$ is said to be a Pareto-optimal solution of the FMLP problem (5.106), if there exists no $x \in \mathbf{R}^n$ such that

$$\mu_{\tilde{X}}(x^p) \leq \mu_{\tilde{X}}(x) \ and \ \mu_{\tilde{F}}(x^p) < \mu_{\tilde{F}}(x),$$
or
$$\mu_{\tilde{X}}(x^p) < \mu_{\tilde{X}}(x) \ and \ \mu_{\tilde{F}}(x^p) \leq \mu_{\tilde{F}}(x).$$

Notice that the Pareto-optimal solution x^p of the FMLP problem (5.106) is a crisp vector. For more properties and relations between Pareto-optimal and compromise solutions, see [135].

Since the problem (5.106) is a maximization problem (i.e., "the higher the values of the criteria the better"), the membership functions $\mu_{\tilde{d}_k}$ of $\tilde{d}_k$ should be increasing, or nondecreasing. For the same reason, the fuzzy relations $\tilde{S}_k$ for comparing $\tilde{f}_k(x, \tilde{c}_k)$ and $\tilde{d}_k$ should be of the "greater or equal" type. Here, we consider $\tilde{S}_k$ as T-fuzzy extensions of the usual binary operation $\geq$, where T is a t-norm.

Formally, in Definitions 5.25 and 5.54, the concepts of feasible solution and compromise solution are similar to each other. Therefore, we can take advantage of the results derived already in the preceding part of this chapter. It can be easily shown that in the case of the single-criteria problem and crisp vector parameters c_1, a_i and b, the max-compromise solution coincides with the optimal solution of the classical LP problem, provided that $\tilde{d}_1$ should be increasing. Moreover, if $\tilde{X}^{*\prime}$ is a compromise solution of FMLP problem (5.106) with the parameters $\tilde{c}_1^\prime$, $\tilde{a}_i^\prime$ and $\tilde{b}^\prime$, and $\tilde{X}^{*\prime\prime}$ is a compromise solution of the FMLP problem with the parameters $\tilde{c}_1^{\prime\prime}$, $\tilde{a}_i^{\prime\prime}$ and $\tilde{b}^{\prime\prime}$ such that for all $i \in \mathcal{M}$, $\tilde{c}_1^\prime \subseteq \tilde{c}_1^{\prime\prime}$, $\tilde{a}_i^\prime \subseteq \tilde{a}_i^{\prime\prime}$ and $\tilde{b}^\prime \subseteq \tilde{b}^{\prime\prime}$, then $\tilde{X}^{*\prime} \subseteq \tilde{X}^{*\prime\prime}$. Particularly, in the case of the crisp optimal solution vector x of the single-criteria LP problem with parameters being crisp numbers, the membership degree of the compromise solution of x (of the associated FMLP problem with fuzzy parameters) is equal to one. This fact enables a natural embedding of the class of (crisp) multicriteria LP problems into the class of FMLP problems.

The next proposition is analogous to Proposition 5.32 for a single-criterion problem.

Proposition 5.56. *Let for all $x \in \mathbb{R}^n$,*

$$\mu_{\tilde{F}_j}(x) = \mu_{\tilde{S}_j}\left(\tilde{f}_j(x, \tilde{c}_j), \tilde{d}_j\right), \quad j \in \mathcal{K},$$

and

$$\mu_{\tilde{X}_i}(x) = \mu_{\tilde{R}_i}\left(\tilde{g}_i(x, \tilde{a}_i), \tilde{b}_i\right), \quad i \in \mathcal{M},$$

be the membership functions of the fuzzy criteria and fuzzy constraints of the FMLP problem (5.106), respectively. Let $G_X = G_F = G = \min$ and let $\tilde{d}_k \in \mathcal{F}(\mathbb{R})$, $k \in \mathcal{K}$, be fuzzy goals.
Then the vector $(t^*, x^*) \in \mathbb{R}^{n+1}$ is an optimal solution of the problem

$$
\begin{aligned}
&maximize\ t \\
&subject\ to\ \mu_{\tilde{F}_j}(x) \geq t, k \in \mathcal{K}, \\
&\qquad\qquad \mu_{\tilde{X}_i}(x) \geq t, i \in \mathcal{M}, x \in \mathbb{R}^n,
\end{aligned}
\tag{5.111}
$$

if and only if x^* is the max-compromise solution of the problem (5.106).

Proof. Let x^* be the max-compromise solution of the problem (5.106) and let

$$t^* = \min\{\mu_{\tilde{X}}(x^*), \mu_{\tilde{F}}(x^*)\},$$

where by (5.108),

$$\mu_{\tilde{F}}(x^*) = \min_{j \in K}\{\mu_{\tilde{S}_j}\left(\tilde{f}_j(x^*, \tilde{c}_j), \tilde{d}_j\right)\}$$

and by (5.17),

$$\mu_{\tilde{X}}(x^*) = \min_{i \in M}\{\mu_{\tilde{R}_i}\left(\tilde{g}_i(x^*, \tilde{a}_i), \tilde{b}_i\right)\}.$$

Then

$$t^* = \max_{x \in \mathbb{R}^n} \min\{\mu_{\tilde{X}}(x), \mu_{\tilde{F}}(x)\},$$

$$\mu_{\tilde{S}_j}\left(\tilde{f}_j(x^*,\tilde{c}_j),\tilde{d}_j\right) \ge t^*, j \in \mathcal{K}$$

and

$$\mu_{\tilde{R}_i}\left(\tilde{g}_i(x^*,\tilde{a}_i),\tilde{b}_i\right) \ge t^*, i \in \mathcal{M}.$$

Consequently, $(t^*,x^*) \in \mathbb{R}^{n+1}$ is an optimal solution of the problem (5.111). On the other hand, let $(t^*,x^*) \in \mathbb{R}^{n+1}$ be an optimal solution of the problem (5.111). Then

$$t^* = \min_{j \in \mathcal{K}, i \in \mathcal{M}}\{\mu_{\tilde{S}_j}\left(\tilde{f}_j(x^*,\tilde{c}_j),\tilde{d}_j\right), \mu_{\tilde{R}_i}\left(\tilde{g}_i(x^*,\tilde{a}_i),\tilde{b}_i\right)\},$$

which means that x^* is the max-compromise solution of the problem (5.106).

□

5.11 Numerical example

Let us study the following problem. An investor has a sum of USD 12 million at the beginning of a monitored term and decides about participation in two investment projects. The length of both projects is three years. Leftover resources in every particular year can be put on time deposit. Returns and costs considered are uncertain and can be formulated as fuzzy numbers. The problem is to find a (nonfuzzy) strategy maximizing quantity of resources at the end of the three-year term. This optimal investment problem can be formulated by the following FLP model,

maximize $\tilde{c}_1 x_1 \tilde{+} \tilde{c}_2 x_2 \tilde{+} (1 \tilde{+} \tilde{u}_3)p_3$

subject to
$$
\begin{array}{llll}
\tilde{a}_{11}x_1 \tilde{+} \tilde{a}_{12}x_2 & \tilde{+}\, p_1 & & \cong 12, \\
\tilde{a}_{21}x_1 \tilde{+} \tilde{a}_{22}x_2 & \tilde{+}(1 \tilde{+} \tilde{u}_1)p_1 & \tilde{-}p_2 & \cong 0, \\
\tilde{a}_{31}x_1 \tilde{+} \tilde{a}_{32}x_2 & & \tilde{+}(1 \tilde{+} \tilde{u}_2)p_2 \tilde{-}p_3 & \cong 0, \\
& x_1, x_2 & & \le 1, \\
& x_1, x_2, p_1, p_2, p_3 & & \ge 0,
\end{array}
$$

(5.112)

where we denote

$\tilde{c}_i$ - fuzzy return of ith project, $i = 1,2,$ at the end of the period;
$\tilde{a}_{ij}$ - fuzzy return/cost of ith proj. $i = 1,2,$ in jth year, $j = 1,2,3$;
$\tilde{u}_j$ - fuzzy interest rate in jth year, $j = 1,2,3$;
x_i - participation measure in ith project, $i = 1,2$;
p_j - resource allocation in jth year, $j = 1,2,3$;
$\cong$ - fuzzy equality relation.

Let $\tilde{a} = (a_L, a_C, a_R)$ be a *triangular fuzzy number*, $a_L < a_C < a_R$, where a_L is called the *left value* of $\tilde{a}$, a_C is called the *central value* and a_R is the *right value* of $\tilde{a}$. Then the membership function of $\tilde{a}$ is given by

$$\mu_{\tilde{a}}(t) = \max\{0, \min\{\frac{t - a_L}{a_C - a_L}, \frac{a_R - t}{a_R - a_C}\}\}. \tag{5.113}$$

If $a_L = a_C = a_R$, we say that $\tilde{a} = (a_L; a_C; a_R)$ is *crisp* (i.e., an ordinary real number) with the membership function identical to the characteristic function χ_{a_C}.

In our problem, the parameters $\tilde{c}_1, \tilde{c}_2, \tilde{a}_{11}, \tilde{a}_{12}, \tilde{a}_{21}, \tilde{a}_{22}, \tilde{a}_{31}, \tilde{a}_{32}, \tilde{u}_1, \tilde{u}_2, \tilde{u}_3$ are supposed to be triangular fuzzy numbers as follows.

$$\begin{aligned}
&\tilde{c}_1 = (4, 6, 8) && \tilde{a}_{21} = (-4, -2, 0) && \tilde{u}_1 = (0.01, 0.02, 0.03) \\
&\tilde{c}_2 = (3, 5, 7) && \tilde{a}_{22} = (1, 2, 3) && \tilde{u}_2 = (0.01, 0.02, 0.03) \\
&\tilde{a}_{11} = (6, 10, 14) && \tilde{a}_{31} = (6, 8, 10) && \tilde{u}_3 = (0.01, 0.03, 0.05) \\
&\tilde{a}_{12} = (3, 6, 9) && \tilde{a}_{32} = (6, 12, 18).
\end{aligned}$$

Let $x_1, x_2 \geq 0$, except $x_1 = x_2 = 0$; i.e., we exclude the situation that the investor will participate in no project.

(a) *Membership functions.* By the extension principle, the left-hand sides of the three constraints in (5.112), denoted by $\tilde{L}_1, \tilde{L}_2, \tilde{L}_3$, are triangular fuzzy numbers as follows.

$$\begin{aligned}
\tilde{L}_1 &= (6x_1 + 3x_2 + p_1, 10x_1 + 6x_2 + p_1, 14x_1 + 9x_2 + p_1), \\
\tilde{L}_2 &= (-4x_1 + x_2 + 1.01p_1 - p_2, -2x_1 + 2x_2 + 1.02p_1 - p_2, \\
&\quad 3x_2 + 1.03p_1 - p_2), \\
\tilde{L}_3 &= (6x_1 + 6x_2 + 1.01p_2 - p_3, 8x_1 + 12x_2 + 1.02p_2 - p_3, \\
&\quad 10x_1 + 18x_2 + 1.03p_2 - p_3).
\end{aligned}$$

Applying (5.113), we calculate the membership functions of $\tilde{L}_1, \tilde{L}_2, \tilde{L}_3$:

$$\begin{aligned}
\mu_{\tilde{L}_1}(t) &= \max\{0, \min\{\frac{t - 6x_1 - 3x_2 - p_1}{4x_1 + 3x_2}, \frac{14x_1 + 9x_2 + p_1 - t}{4x_1 + 3x_2}\}\}, \\
\mu_{\tilde{L}_2}(t) &= \max\{0, \min\{\frac{t + 4x_1 - x_2 - 1.01p_1 + p_2}{2x_1 + x_2 + 0.01p_1}, \frac{3x_2 + 1.03p_1 - p_2 - t}{2x_1 + x_2 + 0.01p_1}\}\}, \\
\mu_{\tilde{L}_3}(t) &= \max\{\min\{0, \frac{t - 6x_1 - 6x_2 - 1.01p_2 + p_3}{2x_1 + 6x_2 + 0.01p_2}, \frac{10x_1 + 18x_2 + 1.03p_2 - p_3 - t}{2x_1 + 6x_2 + 0.01p_2}\}\}.
\end{aligned}$$

Now, we calculate the membership function $\mu_{\cong}$ of the fuzzy relation $\cong$ being a fuzzy extension of the valued relation "=":

$$\mu_{\cong}(\tilde{L}_i, \tilde{P}_i) = \sup\{\min\{0, \mu_{\tilde{L}}(u), \mu_{\tilde{P}_i}(v)\}u = v\}, \quad i = 1, 2, 3,$$

where $\tilde{P}_i$ are crisp numbers given by the membership functions

$$\mu_{\tilde{P}_1}(t) = \begin{cases} 1 & \text{if } t = 12 \\ 0 & \text{otherwise} \end{cases}, \quad \mu_{\tilde{P}_2}(t) = \begin{cases} 1 & \text{if } t = 0 \\ 0 & \text{otherwise} \end{cases},$$

$$\mu_{\tilde{P}_3}(t) = \begin{cases} 1 & \text{if } t = 0 \\ 0 & \text{otherwise} \end{cases}.$$

Particularly,

$$\mu_{\cong}(\tilde{L}_1, \tilde{P}_1) = \mu_{\tilde{L}_1}(12), \quad \mu_{\cong}(\tilde{L}_i, \tilde{P}_i) = \mu_{\tilde{L}_i}(0), \quad i = 2, 3.$$

Notice that for crisp numbers the fuzzy relation "$\cong$" is identical to the ordinary equality relation "$=$".

(b) *Feasible solution.*

By Definition 5.25, using $T^A = T = \min$, the feasible solution of FLP problem (5.112) is a fuzzy set $\tilde{X}$ defined by the membership function:

$$\mu_{\tilde{X}}(x_1, x_2, p_1, p_2, p_3) = \min\{\mu_{\tilde{L}_1}(12), \mu_{\tilde{L}_2}(0), \mu_{\tilde{L}_3}(0)\}.$$

For $\alpha \in (0, 1]$, an α-feasible solution is the set of all vectors

$$x = (x_1, x_2, p_1, p_2, p_3)$$

such that

$$\min\{\mu_{\tilde{L}_1}(12), \mu_{\tilde{L}_2}(0), \mu_{\tilde{L}_3}(0)\} \geq \alpha. \tag{5.114}$$

Inequality (5.114) can be expressed equivalently by the following inequalities.

$$
\begin{aligned}
(6 + 4\alpha)x_1 + (3 + 3\alpha)x_2 + p_1 &\leq 12, \\
(14 - 4\alpha)x_1 + (9 - 3\alpha)x_2 + p_1 &\geq 12, \\
(4 - 2\alpha)x_1 - (1 + \alpha)x_2 - (1.01 + 0.01\alpha)p_1 + p_2 &\geq 0, \\
- 2\alpha x_1 + (3 - \alpha)x_2 + (1.03 - 0.01\alpha)p_1 - p_2 &\geq 0, \\
(6 + 2\alpha)x_1 + (6 + 6\alpha)x_2 + (1.01 + 0.01\alpha)p_2 - p_3 &\leq 0, \\
(10 - 2\alpha)x_1 + (18 - 6\alpha)x_2 + (1.03 - 0.01\alpha)p_2 - p_3 &\geq 0, \\
x_1, x_2 &\leq 1, \\
x_1, x_2, p_1, p_2, p_3 &\geq 0.
\end{aligned}
\tag{5.115}
$$

(c) *Satisficing solution.*

Consider a fuzzy goal $\tilde{d}$ given by the membership function: $\mu_{\tilde{d}}(t) = \min\{1, \max\{0, \frac{t-21}{6}\}\}$ for all $t \geq 0$.

For the membership function of objective $\tilde{Z}$ we have

$$\mu_{\tilde{Z}}(t) = \max\{0, \min\{\frac{t - 4x_1 - 3x_2 - 1.01p_3}{2x_1 + 2x_2 + 0.02p_3}, \frac{8x_1 + 7x_2 + 1.05p_3 - t}{2x_1 + 2x_2 + 0.02p_3}\}\}.$$

Now, we calculate the membership function $\mu_{\succsim}$ of the fuzzy relation $\succsim$ being a fuzzy extension of the valued relation "$\geq$":

$$\mu_{\succsim}(\tilde{Z}, \tilde{d}) = \sup\{\min\{0, \mu_{\tilde{Z}}(u), \mu_{\tilde{d}}(v)\} \mid u \geq v\}.$$

For the membership function of the objective function we obtain

$$\mu_{\geq}(\tilde{Z}, \tilde{d}) = \max\{0, \min\{\frac{8x_1 + 7x_2 + 1.05p_3 - 21}{2x_1 + 2x_2 + 0.02p_3 + 6}, 1\}\}.$$

For the optimal solution $\tilde{X}_0$ it follows that

$$\mu_{\tilde{X}_0}(x) = \min\{\mu_{\tilde{X}}(x_1, x_2, p_1, p_2, p_3), \mu_{\geq}(\tilde{Z}, \tilde{d})\}.$$

For $\alpha \in (0, 1]$, the α-satisficing solution is a set of all vectors $x^0 = (x_1, x_2, p_1, p_2, p_3)$, such that $\mu_{\tilde{X}_0}(x) \geq \alpha$, or, $\mu_{\tilde{X}}(x_1, x_2, p_1, p_2, p_3) \geq \alpha$, and at the same time $\mu_{\geq}(\tilde{Z}, \tilde{d}) \geq \alpha$. The former inequality is equivalent to inequalities (5.115); the latter is equivalent to

$$(8 - 2\alpha)x_1 + (7 - 2\alpha)x_2 + (1.05 - 0,02\alpha)p_3 \geq 21 + 6\alpha. \qquad (5.116)$$

Hence, the set of all α-satisficing solutions is a set of all vectors $x^0 = (x_1, x_2, p_1, p_2, p_3)$ satisfying (5.115) and (5.116).

In order to find a max-satisficing solution of FLP (5.112), we apply Proposition 5.32 by solving the following nonlinear programming problem,

$$\text{maximize } \alpha$$
$$\text{subject to (5.115), (5.116),}$$
$$0 \leq x_1, x_2, \alpha \leq 1,$$
$$p_1, p_2, p_3 \geq 0.$$

By using Excel Solver, we have calculated the following optimal solution.

$$x_1 = 0.605, \ x_2 = 1, \ p_1 = 0, \ p_2 = 0.811, \ p_3 = 17.741, \ \alpha = 0.990.$$

For the crisp problem, i.e., the usual LP problem (5.112), where parameters $\tilde{c}_i, \tilde{a}_{ij}, \tilde{u}_j$ are crisp numbers equal to the central values, we obtain the following optimal solution.

$$x_1 = 0.6, \ x_2 = 1, \ p_1 = 0, \ p_2 = 0.8, \ p_3 = 17.611, \ z = 26.744.$$

Both solutions are close to each other, which is natural, as the central values of the parameters are applied.

On the other hand we could ask for an α-satisficing solution with $\alpha < 1$, e.g., $\alpha = 0.7$, i.e., with a lower level of satisfaction. We have found such a solution with the additional property that p_3 is maximized:

$$x_1 = 0.62, \ x_2 = 1, \ p_1 = 0, \ p_2 = 0.811, \ \alpha = 0.7 \text{ and } p_3 = 17.744.$$

Hence, the fuzzy LP problem formulation allows for finding different kinds of "optimal" solutions in an environment with uncertain parameters of the model, and also enables us to take into account additional requirements.

5.12 Conclusion

In this chapter we have proposed a new general approach to fuzzy single and multicriteria linear programming problems with fuzzy coefficients. A unifying concept of this approach is the concept of a fuzzy relation, particularly fuzzy extension of the inequality or equality relation and the concept of an aggregation operator.

We have formulated the FLP problem, defined a feasible solution of the FLP problem and dealt with the problem of "optimal solution" of FLP problems. Two approaches have been introduced: the satisficing solution based on external goals modeled by fuzzy quantities, and the α-efficient (nondominated) solution. Then our interest has been focused on the problem of duality in FLP. Finally, we have also dealt with the multicriteria case. We have formulated a fuzzy multicriteria linear programming problem, defined a compromise solution and derived basic results. The chapter has been closed with a numerical example.

6

Interval linear systems and optimization problems over max-algebras

K. Zimmermann

6.1 Introduction

In the previous chapters, mostly linear systems of equations and inequalities as well as linear optimization problems with inexact data were investigated. The investigation took advantage of some well-known properties of linear systems and linear problems with exact data. Linear optimization problems are special convex optimization problems, each local extremum value of which (i.e., local maximum or minimum) is at the same time global. In the sequel, we investigate another class of optimization problems, the special structure of which makes it possible to find global optimal solutions. These problems are so-called max-separable optimization problems considered, e.g., in [199]. A special subclass of the max-separable problems form the problems, the objective and constraint functions of which can be treated as "linear" with respect to a pair of operations consisting either of two semigroup operations or of one semigroup and one group operation. We investigate optimization problems with such "linear" objective and constraint functions. The properties of such optimization problems with exact data, which are known from the literature (see, e.g., [29]) are made use of for investigating these problems with interval data. Since the theoretical results for the "linear" optimization problems follow from more general results for the so-called max-separable optimization problems, we derive in the next section these general results and then apply these results to the "linear" problems.

6.2 Max-separable functions and max-separable optimization problems

Definition 6.1. *A function* $f(x_1, \ldots, x_n) = \max\limits_{1 \le n} f_j(x_j)$, *where* f_j *is a real continuous function defined on* $(-\infty, \infty)$, *is called a max-separable function.*

In this section we investigate optimization problems, with a max-separable objective function and max-separable constraints of the following form (such problems are called in the sequel max-separable optimization problems).

$$\text{minimize}\ \ f(x) = \max_{1 \le j \le n} f_j(x_j) \tag{6.1}$$

subject to

$$r_i(x) = \max_{1 \le j \le n} r_{ij}(x_j) \ge b_i \quad i = 1, \ldots, m \tag{6.2}$$

$$r_i(x) = \max_{1 \le j \le n} r_{ij}(x_j) \le b_i \quad i = m+1, \ldots, m_1 \tag{6.3}$$

$$\underline{x}_j \le x_j \le \overline{x}_j \quad\quad j = 1, \ldots, n, \tag{6.4}$$

where for all i, j, $\underline{x}_j$, $\overline{x}_j$, b_i are given finite numbers and r_{ij} are nondecreasing continuous functions defined on $(-\infty, \infty)$.

Remark 6.2. The max-separable optimization problems considered in this section form a special class of nonconvex problems with a nondifferentiable objective function and nondifferentiable functions in the constraints. The special structure of the constraints makes it possible to avoid complications, which might be caused by local extremal values and find the global extremum (or more exactly one of the optimal solutions, in which the global minimum or maximum of the objective function is attained). In the next section, special subclasses of these problems are considered and solved by making use of the general scheme described in the sequel in this section. In these problems, functions $r_{ij}(x_j)$ have a special simple form, namely either $r_{ij}(x_j) = a_{ij} + x_j$ or $r_{ij}(x_j) = \min(a_{ij}, x_j)$, where a_{ij} are given real numbers and x_j are variables. These problems were motivated by some operations research problems (e.g. machine-time scheduling, reliability of networks) and are interesting also from a purely theoretical point of view. The investigation of such problems led to the development of special algebraic structures, so-called extremal algebras, in which the problems with $r_{ij}(x_j) = a_{ij} + x_j$ or $r_{ij}(x_j) = \min(a_{ij}, x_j)$ can be viewed as linear problems with respect to a special concept of linearity, which can be introduced in these structures. The properties of these algebraic structures and optimization problems are thoroughly investigated in the next sections. The main results are proved for general max-separable optimization problems in this section and then, in the next sections, applied to special problems, which are "linear" with respect to the pair of operations introduced in the extremal algebras.

Since for any i $\max_{1 \le j \le n} r_{ij}(x_j) \le b_i$ if and only if $r_{ij}(x_j) \le b_i$ for all j, the constraints (6.3), (6.4) are equivalent to $x_j \le \overline{\overline{x}}_j$, where $\overline{\overline{x}}_j \le \overline{x}_j$, we can assume without loss of generality that the constraints (6.3), (6.4) have already been replaced by the new upper bounds and consider only problems of the form (6.1), (6.2), (6.4). Let

$$V_{ij} = \{x_j \mid r_{ij}(x_j) \geq b_i, \underline{x}_j \leq x_j \leq \overline{x}_j\} \quad \text{for all } i, j. \tag{6.5}$$

Constraint $\max_{1 \leq j \leq n} r_{ij}(x_j) \geq b_i$ can be replaced for any $i = 1, \ldots, m$ by the following equivalent disjunctive constraint,

$$\text{either} \quad x_1 \in V_{i1} \quad \text{or} \quad x_2 \in V_{i2} \quad \text{or} \quad \ldots \quad x_n \in V_{in}. \tag{6.6}$$

Since we assumed that r_{ij}s are nondecreasing continuous functions, either there exists for any $i \in \{1, \ldots, m\}$, $j \in \{1, \ldots, n\}$ $\underline{x}_{ij} \in (-\infty, \infty)$ such that $V_{ij} = \{x_j \mid \underline{x}_j \leq \underline{x}_{ij} \leq x_j \leq \overline{x}_j\}$ or $V_{ij} = \emptyset$. Therefore it holds for any fixed j and two indices $i_1, i_2 \in \{1, \ldots, m\}$ that either $V_{i_1 j} \subseteq V_{i_2 j}$ or $V_{i_2 j} \subseteq V_{i_1 j}$. This property of the constraints (6.2) (or their reformulation (6.6)) is called the "chain property". It follows that if the sets V_{ij} satisfy the chain property, then for any fixed $j \in \{1, \ldots, n\}$, there exists a permutation $\{i_1, \ldots, i_m\}$ of $\{1, \ldots, m\}$ such that

$$V_{i_1 j} \subseteq V_{i_2 j} \subseteq \cdots \subseteq V_{i_m j}. \tag{6.7}$$

We say that the sets V_{ij}, $1 \leq i \leq m$, form a chain.

As a result of these considerations, we obtain that max-separable optimization problems (6.1), (6.2), (6.4) are special cases of the following optimization problems with special disjunctive constraints of the form

$$\text{minimize} \ f(x) = \max_{1 \leq j \leq n} f_j(x_j) \tag{6.8}$$

subject to

$$x_1 \in T_{i1} \text{ or } x_2 \in T_{i2} \quad \text{or} \quad \ldots x_n \in T_{in} \text{ for } i = 1, \ldots, m \tag{6.9}$$

$$\underline{x}_j \leq x_j \leq \overline{x}_j \quad \text{for} \quad j = 1, \ldots, n, \tag{6.10}$$

where T_{ij} are closed subsets of $[\underline{x}_j, \overline{x}_j]$ and satisfy for any fixed $j \in \{1, \ldots, n\}$ the chain property (i.e., the sets $\{T_{ij} \mid 1 \leq i \leq m\}$ form a chain with respect to the set inclusion).

Example 6.3. Let us consider the problem (6.1), (6.2), (6.4) with $r_{ij}(x_j) = a_{ij} + x_j$, where $a_{ij} \in (-\infty, \infty)$ and let

$$T_{ij} = \{x_j \mid \underline{x}_j \leq x_j \leq \overline{x}_j, r_{ij}(x_j) \geq b_i\} = \{x_j \mid \underline{x}_j \leq x_j \leq \overline{x}_j, a_{ij} + x_j \geq b_i\},$$

where $b_i \in (-\infty, \infty)$ for $i = 1, \ldots, m$. Let $\underline{x}_{ij} = \max(b_i - a_{ij}, \underline{x}_j)$. Then we have:

$$T_{ij} = \begin{cases} \emptyset & \text{if } \underline{x}_{ij} > \overline{x}_j \\ [\underline{x}_{ij}, \overline{x}_j] & \text{otherwise.} \end{cases}$$

The sets $\{T_{ij} \mid 1 \leq i \leq m\}$ form a chain and it holds that:

$$T_{i_1 j} \subseteq T_{i_2 j} \quad \text{if and only if} \quad \underline{x}_{i_1 j} \geq x_{i_2 j}.$$

Example 6.4. Let us consider the problem (6.1), (6.2), (6.4) with $r_{ij}(x_j) = \min(a_{ij}, x_j)$, where $a_{ij} \in (-\infty, \infty)$ for all i, j and let

$$T_{ij} = \{x_j \mid \underline{x}_j \leq x_j \leq \overline{x}_j, \min(a_{ij}, x_j) \geq b_i\},$$

where $b_i \in (-\infty, \infty)$. Let $\underline{x}_{ij} = \max\{\underline{x}_j, b_i\}$ for all i, j. Then we have:

$$T_{ij} = \begin{cases} \emptyset & \text{if either } a_{ij} < b_i \text{ or } \overline{x}_j < b_i \\ [\underline{x}_{ij}, \overline{x}_j] & \text{otherwise} \end{cases}$$

and

$$T_{i_1 j} \subseteq T_{i_2 j} \quad \text{if and only if} \quad \underline{x}_{i_1 j} \geq \underline{x}_{i_2 j}.$$

Therefore the sets $\{T_{ij} \mid 1 \leq i \leq m\}$ form for any fixed $j \in \{1, \ldots, n\}$ a chain.

We now investigate some properties of the disjunctive optimization problems (6.8), (6.9), (6.10) under the assumption that the sets T_{ij}, $i = 1, \ldots, m$, $j = 1, \ldots, n$ satisfy the chain property. Such problems were in a more general framework studied in [200].

Theorem 6.5. *The set of feasible solutions of (6.8), (6.9), (6.10) is nonempty if and only if for any $i \in \{1, \ldots, m\}$ there exists an index $q(i)$ such that $T_{iq(i)} \neq \emptyset$.*

Proof. Let M denote the set of feasible solutions of (6.8), (6.9), (6.10) and let us assume that $M \neq \emptyset$. Then there exists for each i at least one index $q(i)$ such that $T_{iq(i)} \neq \emptyset$ (otherwise it would be $T_{ij} = \emptyset$ for all $j = 1, \ldots, n$ and the ith constraint from (6.9) could not be satisfied). Let us assume now that there exists for each i an index $q(i)$, $1 \leq q(i) \leq n$, such that $T_{iq(i)} \neq \emptyset$. Let S_j be defined for each $j \in \{1, \ldots, n\}$ as follows. $S_j = \{i \mid 1 \leq i \leq m, T_{ij} \neq \emptyset\}$. We define further sets T_j, $j \in \{1, \ldots, n\}$ as follows.

$$T_j = \begin{cases} [\underline{x}_j, \overline{x}_j] & \text{if } S_j = \emptyset \\ \bigcap_{i \in S_j} T_{ij} & \text{otherwise,} \end{cases} \tag{6.11}$$

Since we assumed that sets T_{ij}, $i = 1, \ldots, m$ form a chain for any fixed j, the set T_j is equal to one of the sets T_{ij}, $i \in S_j$ whenever $S_j \neq \emptyset$. Therefore T_j is nonempty for all j. According to our assumption, for each i there exists $q(i)$ such that $T_{iq(i)} \neq \emptyset$ so that $i \in S_{q(i)}$ and $T_{q(i)} \subseteq T_{iq(i)}$.

Let us choose $\hat{x}_j \in T_j$ for all $j \in \{1, \ldots, n\}$. We show that $\hat{x}^T = (\hat{x}_1, \ldots, \hat{x}_n) \in M$. Really, for any i, $1 \leq i \leq m$, we obtain that $\hat{x}_{q(i)} \in T_{q(i)} \subseteq T_{iq(i)}$ so that the ith constraint of (6.9) is satisfied and it follows from (6.11) that $\hat{x}_j \in [\underline{x}_j, \overline{x}_j]$ for all $j = 1, \ldots, n$, we obtain that $\hat{x} \in M$ and thus $M \neq \emptyset$. $\square$

Example 6.6. Let us consider the following numerical version of the system (6.2), (6.4). Let the set of feasible solutions M be described by the following system of inequalities.

$$r_1(x) = \max(-1 + x_1, 0 + x_2, 1 + x_3, -7 + x_4) \geq 0$$
$$r_2(x) = \max(-2 + x_1, 3 + x_2, 2 + x_3, -8 + x_4) \geq 0$$
$$r_3(x) = \max(1 + x_1, 5 + x_2, -3 + x_3, -9 + x_4) \geq 0$$
$$0 \leq x_j \leq 5 \quad \text{for} \quad j = 1, 2, \ldots, 4.$$

Let V_{ij} be defined as in (6.5). Then we obtain:

$$V_{11} = [1, 5], \qquad V_{21} = [2, 5], \qquad V_{31} = \emptyset.$$

Similarly we obtain:

$$V_{12} = [0, 5], \qquad V_{22} = [0, 5], \qquad V_{32} = [0, 5] \,;$$
$$V_{13} = [0, 5], \qquad V_{23} = [0, 5], \qquad V_{33} = [3, 5] \,;$$
$$V_{14} = \emptyset, \qquad V_{24} = \emptyset, \qquad V_{34} = \emptyset \,.$$

If we set $T_{ij} = V_{ij}$ for all i, j, our system of constraints can be replaced by the following requirements.

For each $i \in \{1, 2, 3\}$, $x_j \in T_{ij}$ for at least one $j \in \{1, 2, 3, 4\}$ (6.12)
$0 \leq x_j \leq 5$ for all $j \in \{1, 2, 3, 4\}$

If we define T_j as in (6.11), we have $S_1 = \{1, 2\}$, $S_2 = \{1, 2, 3\}$, $S_3 = \{3\}$, $S_4 = \emptyset$ and thus

$$T_1 = [2, 5], \quad T_2 = [0, 5], \quad T_3 = [3, 5], \quad T_4 = [0, 5]$$

It can be easily verified that the necessary and sufficient condition from Theorem 6.5 is satisfied so that $M \neq \emptyset$. If we choose, e.g., $\hat{x}_j = 3$ for $j = 1, 2, 3, 4$, we have $\hat{x}_j \in T_j$ for all j and $\hat{x} \in M$.

We derive in the sequel an explicit formula for the optimal solution of the problem (6.8), (6.9), (6.10). Let us remark that we assumed, when formulating this optimization problem, that the sets T_{ij}, $i \in \{1, \ldots, m\}$ form a chain for any $j \in \{1, \ldots, n\}$ and this assumption is therefore assumed also in the next theorem. Let us introduce the following notations for any i, j ($1 \leq i \leq m$, $1 \leq j \leq n$).

$$\hat{x}_j^{(i)} \in \arg\min\{f_j(x_j) \mid x_j \in T_{ij}\}, \quad \text{if} \quad T_{ij} \neq \emptyset, \tag{6.13}$$
$$\hat{f}_i = \min\{f_j(\hat{x}_j^{(i)}) \mid 1 \leq j \leq n, T_{ij} \neq \emptyset\}, \tag{6.14}$$
$$P_i = \{j \mid 1 \leq j \leq n \quad \text{and} \quad f_j(\hat{x}_j^{(i)}) = \hat{f}_i\}. \tag{6.15}$$

Let us note that (6.13) means that $\hat{x}_j^{(i)}$ is any element from T_{ij}, at which the minimum of f_j on the set T_{ij} is attained so that it holds:

$$f_j(\hat{x}_j^{(i)}) = \min_{x_j \in T_{ij}} f_j(x_j). \tag{6.16}$$

Let us remark that $\hat{x}_j^{(i)}$ is defined only if $T_{ij} \neq \emptyset$.

We now prove the following theorem, which gives us an explicit formula for obtaining the optimal solution of the optimization problem (6.8), (6.9), (6.10).

Theorem 6.7. *Let M be the set of feasible solutions of the optimization problem (6.8), (6.9), (6.10) and let $M \neq \emptyset$. Let the notations (6.13), (6.14), (6.15) be introduced. Let us define W_j, $\hat{T}_j$ as follows.*

$$W_j = \{i \mid j \in P_i\}, \tag{6.17}$$

$$\hat{T}_j = \begin{cases} \bigcap_{i \in W_j} T_{ij} & \text{if } W_j \neq \emptyset \\ [\underline{x}_j, \overline{x}_j] & \text{otherwise.} \end{cases} \tag{6.18}$$

Let $\hat{x}_j \in \arg\min\{f_j(x_j) \mid x_j \in \hat{T}_j\}$ for $j = 1, \ldots, n$. Then $\hat{x}^T = (\hat{x}_1, \ldots, \hat{x}_n)$ is the optimal solution of the problem (6.8), (6.9), (6.10).

Proof. Let $i \in \{1, \ldots, m\}$ be arbitrary and $q(i) \in P_i$. Then $i \in W_{q(i)}$ and $\hat{T}_{q(i)} = \bigcap_{i \in W_{q(i)}} T_{iq(i)} \subseteq T_{iq(i)}$ so that $\hat{x}_{q(i)} \in \hat{T}_{q(i)} \subseteq T_{iq(i)}$. Therefore for any i, $1 \leq i \leq m$, we can find $q(i) \in \{1, \ldots, n\}$ such that $\hat{x}_{q(i)} \in \hat{T}_{iq(i)}$ so that (6.9) is fulfilled. Since it follows from the definition of $\hat{T}_j$ that $\hat{T}_j \subseteq [\underline{x}_j, \overline{x}_j]$ for all j (compare (6.9) and (6.18)), we have $\hat{x}_j \in \hat{T}_j \subseteq [\underline{x}_j, \overline{x}_j]$ and (6.10) is satisfied too. Therefore $\hat{x}$ is a feasible solution of (6.8), (6.9), (6.10) (i.e., $\hat{x} \in M$). It remains to prove that $\hat{x}$ is the optimal solution of this optimization problem, i.e., that $f(\hat{x}) \leq f(y)$ for any $y \in M$.

Let y be an arbitrary element of M and let $f(\hat{x}) = f_p(\hat{x}_p)$ (i.e., $\max_{1 \leq j \leq n} f_j(\hat{x}_j) = f_p(\hat{x}_p)$). Then we have to prove that $f_p(\hat{x}_p) \leq f(y)$. If $W_p = \emptyset$, then $T_p = [\underline{x}_j, \overline{x}_j]$ (see (6.18)) and therefore

$$f_p(\hat{x}_p) = \min\{f_p(x_p) \mid x_p \in [\underline{x}_p, \overline{x}_p]\} \leq f_p(y_p) \leq \max_{1 \leq j \leq n} f_j(y_j) = f(y). \tag{6.19}$$

If $W_p \neq \emptyset$, and $f_p(\hat{x}_p) \leq f_p(y_p)$, it is again $f_p(\hat{x}_p) \leq f(y)$. It remains to investigate the case that $W_p \neq \emptyset$ and at the same time $f_p(\hat{x}_p) > f_p(y_p)$ holds. We show that in this case there exists another index $v \in \{1, \ldots, n\}$ such that $f_p(\hat{x}_p) \leq f_v(y_v)$ so that it will be

$$f_p(\hat{x}_p) \leq f_v(y_v) \leq \max_{1 \leq j \leq n} f_j(y_j) = f(y).$$

Since $W_p \neq \emptyset$ and the sets T_{ip}, $i \in W_p$ form a chain, there exists an index $s \in W_p$ such that

$$\hat{T}_p = \bigcap_{i \in W_p} T_{ip} = T_{sp} . \tag{6.20}$$

Since $f_p(y_p) < f_p(\hat{x}_p) = \min\{f_p(x_p) \mid x_p \in \hat{T}_p\}$, it must be $y_p \notin \hat{T}_p = T_{sp}$. Since $y \in M$, there exists an index $q(s) \in \{1, \dots, n\}$, $q(s) \neq p$, such that $y_{q(s)} \in T_{sq(s)}$. We have in this case

$$f_{q(s)}(y_{q(s)}) \geq \min\{f_{q(s)}(x_{q(s)}) \mid x_{q(s)} \in T_{sq(s)}\} = f_{q(s)}(\hat{x}_{q(s)}) \geq \hat{f}_s . \tag{6.21}$$

Since $s \in W_p$, it is $p \in P_s$. We obtain according to (6.13), (6.14), (6.15) and taking into account that $\hat{T}_p = T_{sp}$, the following equalities.

$$\hat{f}_s = \min\{f_j(\hat{x}_j^{(s)}) \mid 1 \leq j \leq n, T_{sj} \neq \emptyset\} = f_s(\hat{x}_p^{(s)}) = f_p(\hat{x}_p) . \tag{6.22}$$

Therefore according to (6.21), (6.22),

$$f_{q(s)}(y_{q(s)}) \geq \hat{f}_s = f_p(\hat{x}_p) \tag{6.23}$$

and we have:

$$f(\hat{x}) = f_p(\hat{x}_p) \leq f_{q(s)}(y_{q(s)}) \leq \max_{1 \leq j \leq n} f_j(y_j) = f(y) . \tag{6.24}$$

We obtained in all cases that $\hat{x} \in M$, $f(\hat{x}) \leq f(y)$. Since y was an arbitrarily chosen element of M, it follows that $\hat{x}$ is the optimal solution of the problem (6.8), (6.9), (6.10). $\square$

Example 6.8. Let us consider the problem

$$\text{minimize } f(x) = \max(2 + x_1, 3 + x_2, -4 + x_3, x_4)$$
$$\text{subject to } x \in M ,$$

where M is given in the same way as in Example 6.6.

We have according to (6.13), (6.14), (6.15) and taking into account that $T_{ij} = V_{ij}$ for all i, j:

$$\hat{x}_1^{(1)} = 1, \qquad \hat{x}_2^{(1)} = \hat{x}_3^{(1)} = 0;$$

$$f_1(\hat{x}_1^{(1)}) = 3, \qquad f_2(\hat{x}_2^{(1)}) = 3, \qquad f_3(\hat{x}_3^{(1)}) = -4, \qquad T_{14} = \emptyset;$$

$$\hat{x}_1^{(2)} = \hat{x}_2^{(2)} = \hat{x}_3^{(2)} = 0;$$

$$f_1(\hat{x}_1^{(2)}) = 2, \qquad f_2(\hat{x}_2^{(2)}) = 3, \qquad f_3(\hat{x}_3^{(2)}) = -4, \qquad T_2 = \emptyset;$$

$$\hat{x}_1^{(3)} = \hat{x}_2^{(3)} = 0, \qquad \hat{x}_3^{(3)} = 3;$$

$$f_1(\hat{x}_1^{(3)}) = 2, \qquad f_2(\hat{x}_2^{(3)}) = 3, \qquad f_3(\hat{x}_3^{(3)}) = -1, \qquad T_{34} = \emptyset;$$

$$\hat{f}_1 = \min\{3, 3, -4\} = -4;$$

$$\hat{f}_2 = \min\{2, 3, -4\} = -4;$$

$$\hat{f}_3 = \min\{2, 3, -1\} = -1;$$

$$P_1 = P_2 = P_3 = \{3\}.$$

Therefore $W_1 = W_2 = \emptyset$, $W_3 = \{1, 2, 3\}$, $W_4 = \emptyset$ and thus $\hat{T}_1 = \hat{T}_2 = [0, 5]$, $\hat{T}_3 = [3, 5]$, $\hat{T}_4 = [0, 5]$. The optimal solution is therefore $\hat{x}^T = (0, 0, 3, 0)$ with

$$f(\hat{x}) = \max(2, 3, -1, 0) = 3.$$

Let us remark that the procedure based in Theorem 6.7 gives us one optimal solution. The problem can have more optimal solutions. In this example, for $x' = (1, 0, 5, 1)^T$ we have $f(x') = \max(3, 3, 1, 1) = 3$, $x' \in M$ and thus x' is another optimal solution.

Example 6.9. Let us consider the optimization problem

$$\text{minimize } f(x) = \max(2 + x_1, 3 + x_2, -4 + x_3, x_4)$$

subject to

$$r_1(x) = \max(-1 + x_1, 0 + x_2, 1 + x_3, -7 + x_4) = 0$$
$$r_2(x) = \max(-2 + x_1, 3 + x_2, 2 + x_3, -8 + x_4) = 0$$
$$r_3(x) = \max(1 + x_1, 5 + x_2, -3 + x_3, -9 + x_4) = 0$$
$$-6 \le x_j \le 7 \text{ for } j = 1, 2, 3, 4,$$

where $r_1(x), r_2(x), r_3(x)$ are defined as in Example 6.6. The system of equality constraints is equivalent to the system

$$r_i(x) \ge 0 \text{ for } i = 1, 2, 3,$$
$$r_i(x) \le 0 \text{ for } i = 1, 2, 3.$$

The constraints

$$r_i(x) \le 0 \text{ for } i = 1, 2, 3$$
$$-6 \le x_j \le 7 \text{ for } i = 1, 2, 3, 4$$

are equivalent with

$$-6 \leq x_1 \leq -1, \quad -6 \leq x_2 \leq -5, \quad -6 \leq x_3 \leq -2, \quad -6 \leq x_4 \leq 7 \,.$$

Therefore the set of feasible solutions of the original problem with the equality constraints is the set of vectors $x^T = (x_1, x_2, x_3, x_4)$, which satisfy the system

$$r_i(x) \geq 0 \quad \text{for} \quad i = 1, 2, 3$$
$$-6 \leq x_1 \leq -1, \quad -6 \leq x_2 \leq -5, \quad -6 \leq x_3 \leq -2, \quad -6 \leq x_4 \leq 7$$

It can be easily verified that in this case the set of feasible solutions is non-empty if the upper-bound vector $\overline{x}^T = (-1, -5, -2, 7)$ is a feasible solution. Since $r_i(\overline{x}) = 0$ for $i = 1, 2, 3$ and $-6 \leq \overline{x}_j \leq 7$ for $j = 1, 2, 3, 4$, the set of feasible solutions is nonempty. Therefore, we can apply the procedure from Theorem 6.7 to solving this optimization problem. We solve the equivalent problem

minimize $f(x) = \max(2 + x_1, 3 + x_2, -4 + x_3, x_4)$
subject to

$$r_i(x) \geq 0, \quad i = 1, 2, 3 \,,$$
$$-6 \leq x_1 \leq -1, \quad -6 \leq x_2 \leq -5, \quad -6 \leq x_3 \leq -2, \quad -6 \leq x_4 \leq 7 \,.$$

Following the procedure from Theorem 6.7 (see also the preceding Example 6.8) we obtain for

$$V_{ij} = \{x_j \mid r_{ij}(x_j) \geq 0\}, \quad 1 \leq i \leq 3, \quad 1 \leq j \leq n :$$
$$V_{11} = \emptyset, \quad V_{12} = \emptyset, \quad V_{13} = \emptyset, \quad V_{14} = \{7\}$$
$$V_{21} = \emptyset, \quad V_{22} = \emptyset, \quad V_{23} = \{-2\}, \quad V_{24} = \emptyset$$
$$V_{31} = \{-1\}, \quad V_{32} = \{-5\}, \quad V_{33} = \emptyset, \quad V_{34} = \emptyset$$
$$\hat{x}_4 = \arg\min\{f_4(x_4) \mid x_4 \in V_{14}\} = 7$$
$$\hat{x}_3 = \arg\min\{f_3(x_3) \mid x_3 \in V_{23}\} = -6 \,.$$

Furthermore, since $f_1(-1) = 1 < f_2(-5) = -2$, we obtain according to Theorem 6.7 that $q(3) = 1$ and thus

$$\hat{x}_2 = \arg\min\{f_2(x_2) \mid x_2 \in [-6, -5]\}, \quad f_2(-6) = -3$$
$$\hat{x}_1 = \arg\min\{f_1(x_1) \mid x_1 \in V_{31}\}, \quad f_1(-1) = 1$$

so that the optimal value of the objective function is $f(\hat{x}) = \max(1, -3, -6, 7)$ $= 7$.

As we have already mentioned, in the following, we investigate the properties of max-separable systems of equations and inequalities and max-separable optimization problems of the form (6.1)–(6.4) with special functions $r_{ij}(x_j)$, namely either $r_{ij}(x_j) = a_{ij} + x_j$ or $r_{ij}(x_j) = \min(a_{ij}, x_j)$, where a_{ij} are given real numbers. Such problems were motivated in the literature (see, e.g., [29]) by some operations research problems. At the end of this section we give simplified versions of two such problems as motivating examples. The first one comes from the machine-time scheduling and the second concerns the transferability of a network.

Example 6.10. Let us suppose that n operations $1, \ldots, n$, each of them with a given deterministic processing time, are carried out in cycles. Let $x_i(r)$ be the release time of operation i $(i = 1, \ldots, n)$ in cycle r and no preemption be allowed. Therefore if operation i is started at a time $x_i(r)$, it will be finished at time $x_i(r) + p_i$. We assume that the following technological restrictions are given.

(1) $x_i(r) \in [\underline{h}_i, \overline{h}_i]$ for all $i \in N = \{1, \ldots, n\}$, where $\underline{h}_i$, $\overline{h}_i$ are given numbers (i.e., each operation must begin within a prescribed time interval).
(2) Operation i in cycle $r + 1$ can begin only after the operations $j \in N^{(i)}$ have been finished in the preceding cycle r; i.e.,

$$x_i(r + 1) \geq x_j(r) + p_j \quad \text{for all } j \in N^{(i)}, \tag{6.25}$$

where $N^{(i)}$ is a given subset of N.
(3) Each operation in cycle $r + 1$ should begin at the earliest possible time without any delay; i.e., taking into account (6.25), it must be

$$x_i(r + 1) = \max_{j \in N^{(i)}} (x_j(r) + p_j) \quad \text{for all } i \in N. \tag{6.26}$$

We now reformulate the requirements (1), (2), (3) in such a way that the connection with max-separable restrictions (6.2), (6.3) can be easily seen. Let a_{ij} be defined for each $i, j, \in N$ as follows.

$$a_{ij} = \begin{cases} p_j, & \text{if } j \in N^{(i)} \\ -\infty & \text{otherwise}. \end{cases} \tag{6.27}$$

System (6.26) can then be rewritten in the form:

$$x_i(r + 1) = \max_{j \in N^{(i)}} (a_{ij} + x_j(r)) \quad i \in N. \tag{6.28}$$

We can require, e.g., that for a given fixed r, it is necessary that $x_i(r + 1) = b_i$ for all $i \in N$, where b_i are given constants. We require further that the release times $x_j(r)$ are as close as possible to some recommended values $\hat{x}_j$ for all $j \in N$, and the distance of time vectors $x(r)^T = (x_1(r), \ldots, x_n(r))$ and $\hat{x}^T = (\hat{x}_1, \ldots, \hat{x}_n)$ is given by

$$\|x(r) - \hat{x}\| = \max_{j \in N} |x_j(r) - \hat{x}_j| . \tag{6.29}$$

Our problem can now be formulated as follows.

$$\text{minimize } \max_{j \in N} |x_j(r) - \hat{x}_j| \tag{6.30}$$

subject to

$$\max_{j \in N^{(i)}} (a_{ij} + x_j(r)) = b_i, \quad \forall i \in N \tag{6.31}$$

$$\underline{h}_j \leq x_j(r) \leq \overline{h}_j \quad \forall j \in N . \tag{6.32}$$

If we replace each equation in (6.31) with two inequalities and set $f_j(x_j(r)) = |x_j(r) - \hat{x}_j|$ and $r_{ij}(x_j(r)) = a_{ij} + x_j(r)$, we see that the problem (6.30), (6.31), (6.32) becomes an optimization problem of the form (6.1)–(6.4) with variables $x_j(r), j \in N$. The functions $r_{ij}(x_j(r))$ are continuous and increasing and $f_j(x_j(r))$ are continuous so that we can apply to it the procedure described by Theorem 6.7.

Example 6.11. Let us consider a network of roads connecting each of the cities A_i, $i = 1, \ldots, m$ with each of the cities B_j, $j = 1, \ldots, n$; the capacity (transferability) of the route between A_i and B_j is equal to a given positive number a_{ij}. We want to build n roads, each of them connecting B_j with a final delivery point D and a capacity x_j. The capacity (transferability) of the route connecting A_i via B_j with the delivery point D is therefore equal to $r_{ij}(x_j)$ = min (a_{ij}, x_j). The costs connected with building a route with a capacity x_j are given by a continuous function $f_j(x_j)$. We may require, e.g., that the capacities x_j for $j = 1, \ldots, n$ must be chosen in such a way that for each A_i the maximum capacity among the capacities of the routes $A_i B_j D$ is at least equal to a given positive number b_i and at the same time x_j must lie within prescribed bounds $\underline{x}_j$, $\overline{x}_j$ for all $j = 1, \ldots, n$. We want to find capacities x_j, $j = 1, \ldots, n$ minimizing the maximum costs and satisfying the described technological requirements. Such a problem can be reformulated as the following max-separable optimization problem,

$$\text{minimize } \max_{j \in N} f_j(x_j) \tag{6.33}$$

subject to

$$\max_{j \in N^{(i)}} \min(a_{ij}, x_j) \geq b_i, \quad \forall i \in \{1, \ldots, n\} \tag{6.34}$$

$$\underline{x}_j \leq x_j \leq \overline{x}_j \quad \forall j \in \{1, \ldots, n\} . \tag{6.35}$$

6.3 Extremal algebra notation

During the last 40 years special algebraic structures appeared, in which the usual addition and multiplication of the classical linear algebra were replaced by another pair of operations $\oplus$, $\otimes$; in these structures operation $\oplus$ is equal to one of the extremal operations "maximum" or "minimum" and operation $\otimes$ is the addition, multiplication, "minimum" (in the case where $\oplus$ = "maximum") or "maximum" (in the case where $\oplus$ = "minimum"). An appropriately chosen subset of real numbers with such two operations can be found in the literature under various names as, e.g., "extremal algebra" ([163], [193], [198]), max-algebra [29], fuzzy algebra [30], path algebra [45] etc.

The structures were probably independently rediscovered several times by several authors. The aim of this chapter is to investigate the properties of the $(\oplus, \otimes)$-linear system of equations and/or inequalities and $(\oplus, \otimes)$-linear optimization problems over $(\oplus, \otimes)$-algebras with interval entries. The necessary concepts and notations are introduced in the next section. There are in principle only two significantly different types of $(\oplus, \otimes)$-algebras, namely the case in which $\oplus$ is a semigroup and $\otimes$ a group operation, and the case in which $\oplus$ and $\otimes$ are semigroup operations. For this reason, we confine ourselves to two cases, which are usually used as principal representatives of the two types mentioned above, namely $(\oplus, \otimes) = (\max, +)$ for the first type and $(\oplus, \otimes) = (\max, \min)$ for the second type. The corresponding sets of reals with these operations extended where necessary by the elements $\{-\infty\}$, $\{\infty\}$ are denoted $\mathbb{R}_1(\max, +, -\infty)$, $\mathbb{R}_2(\max, \min, -\infty, \infty)$. We use mainly the abbreviated notations $\mathbb{R}_1$, $\mathbb{R}_2$ so that we have

$$\mathbb{R}_1 = \mathbb{R}_1(\max, +, -\infty) \ ,$$

where $\alpha \oplus -\infty = \max(\alpha, -\infty) = \alpha$ and $\alpha \otimes (-\infty) = \alpha + (-\infty) = -\infty$ for all $\alpha \in \mathbb{R}_1$. Similarly

$$\mathbb{R}_2 = \mathbb{R}_2(\max, \min, -\infty, \infty)$$

with $\max(\alpha, -\infty) = \alpha$ and $\min(\alpha, -\infty) = -\infty$ for all $\alpha \in \mathbb{R}_2$.

We refer in the sequel to $\mathbb{R}_1$ as $(\max, +)$-algebra and to $\mathbb{R}_2$ as $(\max, \min)$-algebra. When we speak about any of the sets $\mathbb{R}_1$, $\mathbb{R}_2$, we use the abbreviated notation $\mathbb{R}$. The set of all $m \times n$-matrices with entries from $\mathbb{R}_1$, $\mathbb{R}_2$ or $\mathbb{R}$ are denoted $\mathbb{R}_1^{m \times n}$, $\mathbb{R}_2^{m \times n}$ or $\mathbb{R}^{m \times n}$, respectively.

Let

$$\mathbb{R}_1^n = \underbrace{\mathbb{R}_1 \times \cdots \times \mathbb{R}_1}_{n \text{ times}}, \qquad \mathbb{R}_2^n = \underbrace{\mathbb{R}_2 \times \cdots \times \mathbb{R}_2}_{n \text{ times}} .$$

If $x \in \mathbb{R}^n$ is a column vector with components $x_1, \ldots, x_n$, then $x^T = (x_1, \ldots x_n)$ (i.e., x^T is the transpose of x). We define:

$$x^T \otimes y = (x_1 \otimes y_1) \oplus \cdots \oplus (x_n \otimes y_n) = \sum_{j=1}^{n} {}^{\oplus} (x_j \otimes y_j)$$

for any $x = (x_1, \ldots, x_n)^T$, $y = (y_1, \ldots, y_n)^T \in \mathbb{R}_1^n$ or $\mathbb{R}_2^n$. We have:

$$x^T \otimes y = \sum_{j=1}^{n}{}^{\oplus}(x_j \otimes y_j) = \max_{1 \leq j \leq n}(x_j + y_j) \quad \text{if } x, y \in \mathbb{R}_1^n,$$

$$x^T \otimes y = \sum_{j=1}^{n}{}^{\oplus}(x_j \otimes y_j) = \max_{1 \leq j \leq n} \min(x_j, y_j) \quad \text{if } x, y \in \mathbb{R}_2^n.$$

To simplify the formulae, we also use the notation $\alpha \wedge \beta = \min(\alpha, \beta)$ so that we can also write for $x, y \in \mathbb{R}_2^n$,

$$x^T \otimes y = \max_{1 \leq j \leq n}(x_j \wedge y_j).$$

For an $m \times n$-matrix $A \in \mathbb{R}^{m \times n}$ and for each $i = 1, \ldots, m$, $j = 1, \ldots, n$, we define the product $(A \otimes x) \in \mathbb{R}^m$ as follows.

$$(A \otimes x)_i = \sum_{j=1}^{n}{}^{\oplus}(a_{ij} \otimes x_j) \quad \text{for } i = 1, \ldots, m,$$

$$A \otimes x = ((A \otimes x)_1, \ldots (A \otimes x)_m)^T.$$

It is therefore for $i = 1, \ldots, m$:

$$(A \otimes x)_i = \sum_{j=1}^{n}{}^{\oplus}(a_{ij} \otimes x_j) = \max_{1 \leq j \leq n}(a_{ij} + x_j) \quad \text{if } a_{ij} \in \mathbb{R}_1, \ x_j \in \mathbb{R}_1$$

and

$$(A \otimes x)_i = \sum_{j=1}^{n}{}^{\oplus}(a_{ij} \otimes x_j) = \max_{1 \leq j \leq n}(a_{ij} \wedge x_j) \quad \text{if } a_{ij} \in \mathbb{R}_2, \ x \in \mathbb{R}_2^n.$$

If $A_{i\cdot}$ denotes the ith row of the matrix A, it is obvious that $(A \otimes x)_i = A_{i\cdot} \otimes x$ for $i = 1, \ldots, m$. Similarly if $A \in \mathbb{R}^{m \times n}$, $B \in \mathbb{R}^{n \times k}$, we define the product $A \otimes B \in \mathbb{R}^{m \times k}$ with entries $(A \otimes B)_{ij}$ defined for $i = 1, \ldots, m$, $j = 1, \ldots, k$ as follows.

$$(A \otimes B)_{ij} = \sum_{p=1}^{n}{}^{\oplus}(a_{ip} \otimes b_{pj}) = A_{i\cdot} \otimes B_{\cdot j} ,$$

where $A_{i\cdot}$ denotes the ith row of A and $B_{\cdot j}$ denotes the jth column of B. We have therefore for all $i = 1, \ldots, m$, $j = 1, \ldots, k$:

$$(A \otimes B)_{ij} = \max_{1 \leq p \leq n}(a_{ip} + b_{pj}) \quad \text{if } a_{ip}, b_{pj} \in \mathbb{R}_1$$

and

$$(A \otimes B)_{ij} = \max_{1 \leq p \leq n}(a_{ip} \wedge b_{pj}) \quad \text{if } a_{ip}, b_{pj} \in \mathbb{R}_2.$$

It was proved in the literature ([29], [193], [198]) that the operations $\oplus$, $\otimes$ introduced above satisfy both distributive and associative laws and the formulae with these operations can be processed as in the usual Euclidean space;

furthermore, elements $-\infty$, 0 are neutral elements with respect to $\oplus$, $\otimes$ in $\mathbb{R}_1$ and $-\infty$ and $+\infty$ are neutral elements with respect to $\oplus$, $\otimes$ in $\mathbb{R}_2$; i.e., we have

$$\alpha \oplus -\infty = \max(\alpha, -\infty) = \alpha, \qquad \alpha \otimes 0 = \alpha + 0 = \alpha \qquad \text{for } \alpha \in \mathbb{R}_1$$

and

$$\alpha \oplus -\infty = \max(\alpha, -\infty) = \alpha, \qquad \alpha \otimes (+\infty) = \min(\alpha, +\infty) = \alpha \qquad \text{for } \alpha \in \mathbb{R}_2.$$

The introduction of this notation enables us to describe systems of equations or inequalities using a similar notation and rules as in the classical linear algebra and in this way makes calculations more transparent.

6.4 Noninterval systems of $(\oplus, \otimes)$-linear equations and inequalities

We consider systems of $(\oplus, \otimes)$-linear equations of the form

$$A \otimes x = b ,\tag{6.36}$$

where $A \in \mathbb{R}^{m \times n}$, $b \in \mathbb{R}^m$ are given. The system (6.36) can also be written as follows.

$$(A \otimes x)_i = b_i, \qquad i = 1, \dots, m \tag{6.37}$$

or

$$\sum_{j=1}^{n} {}^{\oplus}(a_{ij} \otimes x_j) = b_i, \qquad i = 1, \dots, m . \tag{6.38}$$

Any x satisfying (6.36) is called a *solution of system* (6.36).

If $\mathbb{R} = \mathbb{R}_1$, (6.38) can be written as

$$\max_{1 \leq j \leq n} (a_{ij} + x_j) = b_i, \qquad i = 1, \dots, m \tag{6.39}$$

and if $\mathbb{R} = \mathbb{R}_2$, then (6.38) has the form

$$\max_{1 \leq j \leq n} (a_{ij} \wedge x_j) = b_i, \qquad i = 1, \dots, m. \tag{6.40}$$

Since $\oplus$ is only a semigroup operation, it is not possible to transfer variables from one side of the equations to the other one. In this chapter, we consider only systems with variables on one side. Systems with variables on both sides are not investigated here. For similar reasons, it is necessary to distinguish two types of systems of $(\oplus, \otimes)$-linear inequalities, namely

$$A \otimes x \leq b \tag{6.41}$$

and

$$A \otimes x \geq b . \tag{6.42}$$

Remark 6.12. Properties of the systems (6.38), (6.41), (6.42) were investigated in several papers in the literature (see, e.g., [29], [193], [198]). We summarize only some of them, which are important for further investigations in this chapter.

Lemma 6.13. *Each system (6.41) has a maximum solution* $x^*(A, b) \in \mathbb{R}^n$ *(i.e., such a solution, that if x is any other solution of (6.41), then $x \leq x^*(A, b)$ holds). We have for all $j = 1, \ldots, n$:*

$$x_j^*(A, b) = \min_{1 \leq i \leq m} (b_i - a_{ij}), \qquad \text{if } \mathbb{R} = \mathbb{R}_1 \tag{6.43}$$

and

$$x_j^*(A, b) = \min_{1 \leq i \leq m} \{b_i \mid a_{ij} > b_i\}, \qquad \text{if } \mathbb{R} = \mathbb{R}_2 . \tag{6.44}$$

Proof. Let us first assume that $\mathbb{R} = \mathbb{R}_1$. Then $A \otimes x \leq b$ means that

$$\max_{1 \leq j \leq n} (a_{ij} + x_j) \leq b_i \quad \forall i = 1, \ldots, m \tag{6.45}$$

and thus

$$a_{ij} + x_j \leq b_i \quad \forall i = 1, \ldots, m, \quad j = 1, \ldots, n . \tag{6.46}$$

It must be therefore

$$x_j \leq b_i - a_{ij} \quad \forall i = 1, \ldots, m ; \tag{6.47}$$

i.e.,

$$x_j \leq x_j^*(A, b) = \min_{1 \leq i \leq m} (b_i - a_{ij}) \quad \forall j = 1, \ldots, n \tag{6.48}$$

and $x^*(A, b)$ is the maximum solution of (6.41).

Let now $\mathbb{R} = \mathbb{R}_2$. Then $A \otimes x \leq b$ implies

$$\max_{1 \leq j \leq n} \min(a_{ij}, x_j) \leq b_i \quad \forall i = 1, \ldots, m \tag{6.49}$$

and thus

$$\min(a_{ij}, x_j) \leq b_i \quad \forall i = 1, \ldots, m , \quad j = 1, \ldots n . \tag{6.50}$$

It must be therefore

$$x_j \leq x_j^*(A, b) = \min_{i, a_{ij} > b_i} b_i \quad \forall j = 1, \ldots, n \tag{6.51}$$

and $x^*(A, b)$ is the maximum solution of (6.41). $\qquad \square$

Definition 6.14. *The element $x^*(A, b)$ is called the principal solution of (6.41).*

It follows immediately from the proof of Lemma 6.13 that the set of solutions of (6.41) always has the maximum element $x^*(A, b)$. If $\mathbb{R} = \mathbb{R}_1$, matrix A has in each row and each column at least one finite element (i.e., an element different from $-\infty$) and all components of b are finite, then $x_j^*(A, b)$ is finite for all $j = 1, \ldots, n$. If $\mathbb{R} = \mathbb{R}_2$, $\{i \mid a_{ij} > b_i\} \neq \emptyset$ for all $j = 1, \ldots, n$ and all components of b are finite, then $x_j^*(A, b)$ is finite for all $j = 1, \ldots, n$. We assume in the sequel that the matrix A and the vector b fulfill the conditions, which ensure that $x^*(A, b)$ is finite. Let us remark that it follows from the proof of Lemma 6.13 that the set of solutions of (6.41) is always nonempty and $x^*(A, b)$ is one of its solutions.

The following lemma holds ([24]).

Lemma 6.15. *Let* A^1, A^2 *be two matrices over* $\mathbb{R}$ *and* b^1, b^2 *vectors of appropriate dimension. A system of inequalities over* $\mathbb{R}$

$$A^1 \otimes x \leq b^1 \tag{6.52}$$
$$A^2 \otimes x \geq b^2 \tag{6.53}$$

has a solution if and only if the principal solution $x^*(A^1, b^1)$ *fulfills (6.53).*

Proof. If $A^2 \otimes x^*(A^1, b^1) \geq b^2$, the system (6.52), (6.53) has a solution. For the converse direction, let us suppose that y is a solution of (6.52), (6.53). $A^1 \otimes y \leq b^1$ implies that $y \leq x^*(A^1, b^1)$ and $x^*(A^1, b^1)$ as the principal solution of (6.52) fulfills (6.52). Furthermore, we have $A^2 \otimes x^*(A^1, b^1) \geq A^2 \otimes y \geq b^2$ and therefore $x^*(A^1, b^1)$ also fulfills (6.53). $\qquad\square$

Corollary 6.16. *A system of equations (6.36) has a solution if and only if* $x^*(A, b)$ *fulfills (6.36).*

Proof. The system (6.36) is obviously equivalent to the system $A \otimes x \leq b$, $A \otimes x \geq b$. The assertion then follows immediately from Lemma 6.15 with $A^1 = A^2 = A$, $b^1 = b^2 = b$. $\qquad\square$

In general the set of solutions of (6.36) may be empty. The emptiness of this set depends on the structure of the matrix A and the vector b. We summarize here some facts, which were proved in the literature (see, e.g., [29], [193], [198]).

Let us introduce the following notations.

$$S = \{1, \ldots, m\},$$
$$S_j(x_j) = \{i \mid 1 \leq i \leq m \text{ and } a_{ij} \otimes x_j = b_i\} \qquad \forall j = 1, \ldots, n, \ x_j \in \mathbb{R}.$$

It holds then

Lemma 6.17. *Let* A, b *be a* $m \times n$*-matrix and* $b \in \mathbb{R}^m$, *respectively.*

(a) $x \in \mathbb{R}$ *is a solution of (6.36) if and only if* $x \leq x^*(A, b)$ *and*

$$S = \bigcup_{j=1}^{n} S_j(x_j). \tag{6.54}$$

(b) If $x = (x_1, \ldots, x_n)^T$ is a solution of (6.36), then

$$S_j(x_j) \neq \emptyset \implies x_j = x_j^*(A, b)$$

for all $j = 1, \ldots, n$.

Proof. (a) If $x \leq x^*(A, b)$ and (6.54) is fulfilled, it means that for any $i \in S$ we have

$$a_{ij} \otimes x_j \leq a_{ij} \otimes x_j^*(A, b) \leq b_i \quad \text{for all} \quad j = 1, \ldots, n$$

and there exists $j(i)$ such that $i \in S_{j(i)}(x_{j(i)})$ and thus $a_{ij(i)} \otimes x_{j(i)} = b_i$; therefore

$$\sum_{j=1}^{n} {}^{\oplus} a_{ij} \otimes x_j = a_{ij(i)} \otimes x_{j(i)} = b_i \, ,$$

so that $x = (x_1, \ldots x_n)^T$ is a solution of (6.36). If $x = (x_1, \ldots, x_n)^T$ is a solution of (6.36), it must fulfill the inequality $A \otimes x \leq b$ and therefore also the inequality $x \leq x^*(A, b)$. If $i \notin \bigcup_{j=1}^{n} S_j(x_j)$ (i.e., $i \notin S_j(x_j)$ for all $j = 1, \ldots, n$), then it must be $a_{ij} \otimes x_j < a_{ij} \otimes x_j^*(A, b) \leq b_i$ for all $j = 1, \ldots, n$ and thus $\sum_{j=1}^{n} {}^{\oplus} a_{ij} \otimes x_j < b_i$; therefore x is not a solution of (6.36). $\qquad \square$

(b) If x is a solution of (6.36), it is $x_j \leq x_j^*(A, b)$ for all j; if $x_j < x_j^*(A, b)$, then $a_{ij} \otimes x_j < a_{ij} \otimes x_j^*(A, b) \leq b_i$ for all $i = 1, \ldots, m$. Therefore $S_j(x_j) = \emptyset$. $\qquad \square$

The following consequences are implied immediately by Lemma 6.17.

(i) System (6.36) has no solution if and only if $\bigcup_{j=1}^{n} S_j(x_j^*(A, b)) \neq S$.

(ii) Let $c_{ij} = b_i - a_{ij}$ for $i = 1, \ldots, m$, $j = 1, \ldots, n$ and $\mathbb{R} = \mathbb{R}_1$; then

$$S_j(x_j^*(A, b)) = \left\{ s \mid c_{sj} = \min_{i \in S} c_{ij} \right\} .$$

(iii) Let $\mathbb{R} = \mathbb{R}_2$ and $L_j = \{ i \mid i \in S, a_{ij} > b_i \}$ for all $j = 1, \ldots, n$; then

$$S_j(x_j^*(A, b)) = \left\{ s \mid b_s = \min_{i \in L_j} b_i \right\} .$$

Consequences (i), (ii), (iii) make possible an effective algorithmic verification of the existence of solutions of a system (6.36) and finding the corresponding principal solution.

6.5 Noninterval max-separable optimization problems with $(\oplus, \otimes)$-linear constraints

Let $f_j : \mathbb{R} \to \mathbb{R}$ be continuous functions for $j = 1,\ldots,n$ and $f(x) = \max\limits_{1\leq j\leq n} f_j(x_j)$ for any $x \in \mathbb{R}^n$, so that $f : \mathbb{R}^n \to \mathbb{R}$ is a max-separable function. We consider optimization problems with a max-separable objective function and $(\oplus, \otimes)$-linear functions in the constraints of the following form (we assume $x \in \mathbb{R}^n$).

$$\text{minimize } f(x) \tag{6.55}$$

subject to

$$A \otimes x = b \tag{6.56}$$

$$x_j \geq \underline{x}_j > -\infty \quad \forall j = 1,\ldots,n . \tag{6.57}$$

Remark 6.18. If $f_j(x_j) = c_j \otimes x_j$ for all $j \in N$, the problem considered above can be called a $(\oplus, \otimes)$-linear optimization problem. Since the methods suggested here for solving optimization problems with $(\oplus, \otimes)$-linear constraints with a max-separable objective function are not substantially influenced by the form of the functions f_j in the objective function, we first consider the solution of the problems with a general max-separable objective function and then show how to adjust the methods for the case where the objective function is $(\oplus, \otimes)$-linear. Such an approach is chosen for the noninterval problems in this section as well as for the interval problems in the sequel.

We can assume without the loss of generality in the sequel in this section that the set of feasible solutions of (6.56) is nonempty, since the nonemptiness can be easily verified using (i), (ii), (iii) from the preceding section taking into account (6.57). For the same reason, we can assume that we have at our disposal the principal solution $x^*(A, b)$ of (6.56). To avoid unnecessary complications we assume that all a_{ij}, b_i are finite so that $x_j^*(A, b) \in (-\infty, \infty)$.

Remark 6.19. We consider only minimization problems, since the minimization problems are the most interesting from the point of view of applications (see the next section). As far as the maximization is concerned, let us mention that $x^*(A, b)$ is the optimal solution for any maximization problem with the constraints (6.56) and objective function satisfying the condition:

$$x \leq y \Rightarrow f(x) \leq f(y) \quad \text{for all } x, y \in \mathbb{R}^n .$$

Several procedures were suggested for solving (6.55), (6.56) in the literature (see [198], [199]). We demonstrate here one procedure following from [199], [200]. Let us denote for $i = 1,\ldots,m$, $j = 1,\ldots,n$:

$$V_{ij} = \{x_j \mid a_{ij} \otimes x_j = b_i \text{ and } x_j \leq x_j(A^*, b)\} \tag{6.58}$$

Then we have:

$$V_{ij} = \begin{cases} \emptyset & \text{if } a_{ij} \otimes x_j^*(A, b) < b_i, \\ \{x_j^*(A, b)\} & \text{otherwise}. \end{cases} \tag{6.59}$$

Following the procedure from Theorem 6.7, we determine indices $k(i)$ for $i \in S$ as follows,

$$f_{k(i)}(x_{k(i)}^*(A, b)) = \min_{j, V_{ij} \neq \emptyset} f_j(x_j^*(A, b)) \tag{6.60}$$

and set

$$P_j = \{i \mid i \in S, k(i) = j\}. \tag{6.61}$$

Let us define x_j^{opt} for $j = 1, \ldots, n$ as follows.

$$x_j^{\text{opt}} = \begin{cases} x_j^*(A, b) & \text{if } P_j \neq \emptyset, \\ \arg\min \{f_j(x_j) \mid x_j \in [\underline{x}_j, x_j^*(A, b)]\} & \text{otherwise}. \end{cases} \tag{6.62}$$

It follows from 6.7 that $x^{\text{opt}} = (x_1^{\text{opt}}, \ldots, x_n^{\text{opt}})^T$ is the optimal solution of (6.55)–(6.57).

Remark 6.20. Since f_j are continuous, $\underline{x}_j > -\infty$ and $x_j^*(A, b) < \infty$, the $\min\{f_j(x_j) \mid x_j \in [\underline{x}_j, x_j^*(A, b)]\}$ always exists.

If $f_j(x_j) = c_j \otimes x_j$, then optimization problem (6.55), (6.56) is called a $(\oplus, \otimes)$-linear optimization problem and its optimal solution $\tilde{x}^{\text{opt}}$ is the following (compare (6.62)).

$$\tilde{x}_j^{\text{opt}} = \begin{cases} x_j^*(A, b) & \text{if } P_j \neq \emptyset \\ \underline{x}_j & \text{otherwise} \end{cases} \quad \forall j = 1, \ldots, n. \tag{6.63}$$

We investigate further optimization problems with the objective function $f(x)$ and $(\oplus, \otimes)$-linear inequality constraints:

$$\text{minimize } f(x) = \max_{1 \le j \le n} f_j(x_j) \tag{6.64}$$

subject to

$$A^1 \otimes x \le b^1 \tag{6.65}$$

$$A^2 \otimes x \ge b^2 \tag{6.66}$$

$$x \ge \underline{x}, \tag{6.67}$$

where A^1 is a $m_1 \times n$-matrix and A^2 a $m_2 \times n$-matrix over $\mathbb{R}$. We again assume that $\underline{x}_j > -\infty \ \forall j$ and $x_j^*(A^1, b^1) < \infty$ for all $j = 1, \ldots, n$. It can be easily shown that under these assumptions the set of feasible solutions of this problem is a compact set. Since $f(x)$ is under our assumptions continuous, the optimal solution of (6.64)–(6.67) exists, whenever the set of feasible solutions of this problem is nonempty. We present a procedure for solving (6.64)–(6.67), which follows again from [199], [200].

Let us denote for all $j = 1, \ldots, n$, $i = 1, \ldots, m$:

$$W_{ij} = \{x_j \mid x_j \in [\underline{x}_j, x_j^*(A^1, b^1)] \text{ and } a_{ij}^2 \otimes x_j \ge b_i^2\}. \tag{6.68}$$

Lemma 6.21. *The set of feasible solutions of (6.64)–(6.67) is empty if and only if there exists $i_0 \in \{1, \ldots, m_2\}$ such that $W_{i_0 j} = \emptyset$ for all $j = 1, \ldots, n$.*

Proof. If $W_{i_0 j} = \emptyset$ for all $j = 1, \ldots, n$, then for any $x_j \in [\underline{x}_j, x_j^*(A^1, b^1)]$

$$a_{i_0 j}^2 \otimes x_j < b_{i_0}^2 \text{ for all } j \text{ and thus } \sum_{j=1}^{n} {}^{\oplus} a_{i_0 j}^2 \otimes x_j < b_{i_0} \text{ for any } x \text{ satisfying (6.65)}.$$

Therefore (6.66) is not fulfilled and the set of feasible solutions of (6.64)–(6.67) is empty. Let on the contrary for each $i \in \{1, \ldots, m_2\}$ there exist $j(i) \in \{1, \ldots, n\}$ such that $W_{ij(i)} \neq \emptyset$. Let us note that if for any $j \in \{1, \ldots, n\}$, sets $W_{i_1 j}, W_{i_2 j} \ i_1, i_2 \in \{1, \ldots, m_2\}$ are nonempty, then it holds always either $W_{i_1 j} \subseteq W_{i_2 j}$ or $W_{i_2 j} \subseteq W_{i_1 j}$. Really, if $\mathbb{R} = \mathbb{R}_1$, it is $W_{i_1 j} \subseteq W_{i_2 j}$, if and only if either $W_{i_1 j} = \emptyset$ or $b_{i_1} - a_{i_1 j} \geq b_{i_2} - a_{i_2 j}$. If $\mathbb{R} = \mathbb{R}_2$, then $W_{i_1 j} \subseteq W_{i_2 j}$ if and only if either $W_{i_1 j} = \emptyset$ or $a_{i_2 j} = b_{i_2}$ and $b_{i_1} \geq b_{i_2}$. Let $P_j = \{i \mid 1 \leq i \leq m_2, j(i) = j\}$. Then if $P_j \neq \emptyset$ there exists $i(j) \in \{1, \ldots, m_2\}$,

$$\bigcap_{i \in P_j} W_{ij(i)} = W_{i(j)j(i(j))} = W_{i(j)j} .$$

Let us define $\tilde{x} = (\tilde{x}_1, \ldots, \tilde{x}_n)^T$ as follows.

$$\tilde{x}_j \in W_{i(j)j} \quad \text{if } P_j \neq \emptyset$$
$$\tilde{x}_j \in [\underline{x}_j, x_j^*(A^1, b^1)] \quad \text{if } P_j = \emptyset .$$

Let $i_0 \in \{1, \ldots, m_2\}$ be arbitrary and $W_{i_0 j(i_0)} = W_{i_0 k} \neq \emptyset$; it holds that $i_0 \in P_k$, so that

$$\tilde{x}_k \in W_{i(k)k} = \bigcap_{i \in P_k} W_{ik} \subseteq W_{i_0 k} = W_{i_0 j(i_0)} .$$

Therefore

$$\sum_{j=1}^{n} {}^{\oplus} a_{i_0 j}^2 \otimes \tilde{x}_j \geq a_{i_0 k} \otimes \tilde{x}_k \geq b_{i_0}^2 ;$$

since i_0 was arbitrarily chosen, it follows that $A^2 \otimes \tilde{x} \geq b^2$; since $\underline{x} \leq \tilde{x} \leq x^*(A^1, b^1)$, it is also $A^1 \otimes \tilde{x} \leq b^1$, $\tilde{x} \geq \underline{x}$ so that relations (6.65)–(6.67) are fulfilled and $\tilde{x}$ is a feasible solution of (6.64)–(6.67). $\square$

We assume in the sequel that the set of feasible solutions of (6.64)–(6.67) is nonempty and describe a procedure following from Theorem 6.7 for determining the optimal solution of (6.64)–(6.67). Let us note that the nonemptiness of the set of feasible solutions of (6.64)–(6.67) can be easily verified by making use of Lemma 6.21.

Let $k(i) \in \{1, \ldots, n\}$ be defined for $i = 1, \ldots, m_2$ as follows.

$$x_j^{(i)} = \arg\min\{f_j(x_j) \mid x_j \in W_{ij}\} \text{ for all } i, j \text{ with } W_{ij} \neq \emptyset \qquad (6.69)$$

$$f_{j(i)}(x_{k(i)}^{(i)}) = \min_{1 \leq j \leq n} f_j(x_j^{(i)}) . \tag{6.70}$$

Let

$$T_j = \{i \mid 1 \leq i \leq m_2 \text{ and } k(i) = j\} \tag{6.71}$$

and

$$W_j = \bigcap_{i \in T_j} W_{ij} = W_{i(j)j}, \quad \text{if } T_j \neq \emptyset . \tag{6.72}$$

Let us define $\tilde{x}^{\text{opt}} = \left(\tilde{x}_1^{\text{opt}}, \ldots, \tilde{x}_n^{\text{opt}}\right)^T$ as follows.

$$\tilde{x}_j^{\text{opt}} = \begin{cases} \arg\min\{f_j(x_j) \mid x_j \in W_{i(j)j}\} & \text{if } T_j \neq \emptyset \\ \arg\min\{f_j(x_j) \mid x_j \in [\underline{x}_j, x_j^*(A^1, b^1)]\} & \text{otherwise.} \end{cases} \tag{6.73}$$

Then $\tilde{x}^{\text{opt}} = \left(\tilde{x}_1^{\text{opt}}, \ldots, \tilde{x}_n^{\text{opt}}\right)^T$ is the optimal solution of (6.64)–(6.67).

If $f_j(x_j) = c_j \otimes x_j$ for all j, we obtain the so-called ($\oplus, \otimes$)-linear optimization problem with ($\oplus, \otimes$)-linear inequality constraints. It can be easily verified that since f_j is nondecreasing, we have:

$$\tilde{x}_j^{\text{opt}} = \begin{cases} \inf(W_{i(j)j}), & \text{if } T_j \neq \emptyset \\ \underline{x}_j & \text{otherwise} \end{cases} \quad \forall j = 1, \ldots, n , \tag{6.74}$$

where $\inf(W_{i(j)j})$ is the minimum element of $W_{i(j)j}$.

Remark 6.22. (6.62) follows from (6.72), if we set in (6.64)–(6.67) $A^1 = A^2 = A$, $b^1 = b^2 = b$.

In the following, we use the obtained results to derive new results for ($\oplus, \otimes$)-linear systems of equations and inequalities and ($\oplus, \otimes$)-linear optimization problems with interval data. Some of the results and concepts can be found in [25], [24], [193].

6.6 ($\oplus, \otimes$)-linear systems of equalities and inequalities with interval coefficients

Let $\underline{A}, \overline{A}$ be two matrices from $\mathbb{R}^{m \times n}$, $\underline{A} \leq \overline{A}$. We define

$$\mathbf{A} = [\underline{A}, \overline{A}] = \{A \mid \underline{A} \leq A \leq \overline{A}\} . \tag{6.75}$$

Similarly for $\underline{b}, \overline{b} \in \mathbb{R}^m$, $\underline{b} \leq \overline{b}$ we define

$$\mathbf{b} = [\underline{b}, \overline{b}] = \{b \mid \underline{b} \leq b \leq \overline{b}\} . \tag{6.76}$$

The interval system of ($\oplus, \otimes$)-linear equations and/or inequalities is a system of the form

$$(\mathbf{A} \otimes x)_i \sim_i \mathbf{b}_i, \quad i = 1, \ldots, k , \tag{6.76}$$

where $\sim_i$ is for all i one of the relations $\leq, =, \geq$.

Definition 6.23. *We say that the system (6.76) is weakly solvable if there exist $A \in \mathbf{A}$ and $b \in \mathbf{b}$ such that $(A \otimes x)_i \sim_i b_i$, $i = 1, \ldots, m$ is solvable. The system is called strongly solvable if $(A \otimes x)_i \sim_i b_i$ is solvable for all $A \in \mathbf{A}$, $b \in \mathbf{b}$.*

Definition 6.24. *A vector $x \in \mathbb{R}^n$ is called*

(1) A weak solution of (6.76), if there exist $A \in \mathbf{A}$ and $b \in \mathbf{b}$ such that $(A \otimes x)_i \sim_i b_i$ for all i;

(2) A tolerance solution of (6.76), if for each $A \in \mathbf{A}$, $A \otimes x \in \mathbf{b}$;

(3) A strong solution, if for each $A \in \mathbf{A}$, and each $b \in \mathbf{b}$ $(A \otimes x)_i \sim_i b_i$ for all i holds.

Let us consider an interval system of equations of the form

$$A \otimes x = b , \tag{6.77}$$

where $\mathbf{A} \subseteq \mathbb{R}^{m \times n}$, $\mathbf{b} \subseteq \mathbb{R}^m$ (i.e., $\sim_i$ are equalities $\forall i$ in (6.76)).

Theorem 6.25 ([25]). *A vector x is a weak solution of the interval system (6.77) if and only if $\underline{A} \otimes x \leq \overline{b}$ and $\overline{A} \otimes x \geq \underline{b}$.*

Proof. Let $i = \{1, \ldots, m\}$ be an arbitrarily chosen index and $x = (x_1, \ldots, x_n)^T \in \mathbb{R}^n$ fixed; let $f_x^{(i)}(a_{i1}, \ldots, a_{in})$ be a function of n variables defined as follows.

$$f_x^{(i)}(a_{i1}, \ldots, a_{in}) = \sum_{j=1}^n {}^\oplus (a_{ij} \otimes x_j) = (A \otimes x)_i . \tag{6.78}$$

If $A \in \mathbf{A}$, then since $f_x^{(i)}(a_{i1}, \ldots, a_{in})$ is isotone, we have:

$$f_x^{(i)}(a_{i1}, \ldots, a_{in}) = (A \otimes x)_i \in I_i = [(\underline{A} \otimes x)_i, (\overline{A} \otimes x)_i] \subseteq \mathbb{R} . \tag{6.79}$$

Hence x is a weak solution if and only if

$$(A \otimes x)_i \in I_i \cap [\underline{b}_i, \overline{b}_i] \quad \forall i = 1, \ldots, m . \tag{6.80}$$

Therefore x is a weak solution of (6.77) if and only if $I_i \cap [\underline{b}_i, \overline{b}_i] \neq \emptyset$ for all $i = 1, \ldots, m$; let $b_i \in I_i \cap [\underline{b}_i, \overline{b}_i]$ for all i. Since $f_x^{(i)}(a_{i1}, \ldots, a_{in})$ is for each i a continuous function, there exist $(a_{i1}(b_i), \ldots, a_{in}(b_i))$ such that

$$\sum_{j=1}^n {}^\oplus a_{ij}(b_i) \otimes x_j = b_i \quad \forall i = 1, \ldots, m , \tag{6.81}$$

i.e., $(a_{i1}(b_i), \ldots, a_{in}(b_i))$ is the preimage of b_i for the mapping $f_x^{(i)}$. Let $A(b)$ be the matrix with entries $a_{ij}(b_i)$ from (6.81). Then $A(b) \in \mathbf{A}$, $A(b) \otimes x = b$ and $b \in \mathbf{b}$, which completes the proof.   □

Corollary 6.26. *An interval system (6.78) has a weak solution if and only if* $\overline{A} \otimes x^*(\underline{A}, \overline{b}) \geq \underline{b}$.

Proof. The assertion follows from Theorem 6.25 and Lemma 6.15, if we set in Lemma 6.15 $A^1 = \underline{A}$, $b^1 = \underline{b}$, $A^2 = \overline{A}$, $b^2 = \underline{b}$.   □

Theorem 6.27 ([24]). *A vector x is a tolerance solution of (6.77) if and only if*

$$\underline{A} \otimes x \geq \underline{b} , \tag{6.82}$$
$$\overline{A} \otimes x \leq \overline{b} . \tag{6.83}$$

Proof. If x is a tolerance solution, then $\underline{b} \leq A \otimes x \leq \overline{b}$ for all $A \in \mathbf{A}$; if $A = \underline{A}$ we obtain (6.82) and if $A = \overline{A}$, we obtain (6.83).

For the opposite implication, let us suppose that x fulfills (6.82), (6.83), but it is not a tolerance solution of (6.77); i.e., there exist $\tilde{A} \in \mathbf{A}$, $\tilde{b} \in \mathbf{b}$ and $i_0 \in \{1, \ldots, m\}$ such that either

$$\sum_{j=1}^{n} {}^{\oplus} \tilde{a}_{i_0 j} \otimes x_j < \underline{b}_{i_0} \tag{6.84}$$

or

$$\sum_{j=1}^{n} {}^{\oplus} \tilde{a}_{i_0 j} \otimes x_j > \overline{b}_{i_0} . \tag{6.85}$$

If (6.84) holds, then it contradicts (6.82) and if (6.85) holds, it contradicts (6.83). Therefore if x satisfies (6.82), (6.83), it is a tolerance solution of (6.77).   □

Corollary 6.28. *A vector x is a tolerance solution of (6.77) if and only if* $\underline{A} \otimes x^*(\overline{A}, \overline{b}) \geq \underline{b}$.

Proof. The assertion follows immediately from Theorem 6.27, if we set in Lemma 6.15 $A^1 = \overline{A}$, $b^1 = \overline{b}$, $A^2 = \underline{A}$, $b^2 = \underline{b}$.   ■

Theorem 6.29 ([24]). *A vector x is a strong solution of (6.77) if and only if it is a solution of the system*

$$\overline{A} \otimes x = \underline{b} , \tag{6.86}$$
$$\underline{A} \otimes x = \overline{b} . \tag{6.87}$$

Proof. If x is a strong solution of (6.77), it obviously fulfills (6.86), (6.87). Conversely, let x fulfill (6.86), (6.87) and let $\tilde{A} \in \mathbf{A}$, $\tilde{b} \in \mathbf{A}$ exist such that $\tilde{A} \otimes x \neq \tilde{b}$. Then, there exists an index $i_0 \in \{1, \ldots, m\}$ such that either $(\tilde{A} \otimes x)_{i_0} < \tilde{b}_{i_0}$ or $(\tilde{A} \otimes x)_{i_0} > \tilde{b}_{i_0}$. In the former case. it is $(\underline{A} \otimes x)_{i_0} \leq (\tilde{A} \otimes x)_{i_0} < \tilde{b}_{i_0}$ so that (6.87) is violated and in the latter case we have $(\overline{A} \otimes x)_{i_0} \geq (\tilde{A} \otimes x)_{i_0} > \tilde{b}_{i_0}$; i.e., (6.86) is not fulfilled.   □

Corollary 6.30. *An interval system (6.77) has a strong solution if and only if $\overline{A} \otimes x^* \geq \underline{b}$ and $\underline{A} \otimes x^* \geq \overline{b}$, where $x^* = x^* \left(\begin{pmatrix} \overline{A} \\ \underline{A} \end{pmatrix}, \begin{pmatrix} \underline{b} \\ \overline{b} \end{pmatrix} \right)$ is the principal solution of the inequality system*

$$\overline{A} \otimes x \leq \underline{b}, \tag{6.88}$$
$$\underline{A} \otimes x \leq \overline{b}. \tag{6.89}$$

Remark 6.31. Note that if x^* in the corollary of Theorem 6.29 satisfies (6.88), (6.89), then x^* is a strong solution of (6.77).

Let us consider further an interval system of inequalities of the form

$$\mathbf{A} \otimes x \leq \mathbf{b} . \tag{6.90}$$

Theorem 6.32. *A vector x is a weak solution of (6.90) if and only if it is a solution of the system $\underline{A} \otimes x \leq \overline{b}$.*

Proof. If $\underline{A} \otimes x \leq \overline{b}$, then x is obviously a weak solution of (6.90). Conversely, let x be a weak solution of (6.90), i.e., there exist $A \in \mathbf{A}$, $b \in \mathbf{b}$ such that $A \otimes x \leq b$; then we have

$$\underline{A} \otimes x \leq A \otimes x \leq b \leq \overline{b} .$$

$\square$

Let us consider an interval system of inequalities of the form

$$\mathbf{A} \otimes x \geq \mathbf{b} . \tag{6.91}$$

Theorem 6.33. *A vector x is a weak solution of (6.91) if and only if $\overline{A} \otimes x \geq \underline{b}$.*

Proof. If $\overline{A} \otimes x \geq \underline{b}$, then x is obviously a weak solution of (6.91). Conversely, if x is a weak solution of (6.91), i.e., there exist $A \in \mathbf{A}$, $b \in \mathbf{b}$ such that $A \otimes x \geq b$, then we have

$$\overline{A} \otimes x \geq A \otimes x \geq b \geq \underline{b} .$$

$\square$

Let $A^1 \in \mathbb{R}^{m_1 \times n}$, $A^2 \in \mathbb{R}^{m_2 \times n}$,

$$\mathbf{A}^1 = [\underline{A}^1, \overline{A}^1] = \{ A^1 \in \mathbb{R}^{m_1 \times n} \mid \underline{A}^1 \leq A \leq \overline{A}^1 \},$$
$$\mathbf{A}^2 = [\underline{A}^2, \overline{A}^2] = \{ A^2 \in \mathbb{R}^{m_2 \times n} \mid \underline{A}^2 \leq A^2 \leq \overline{A}^2 \} .$$

Let us consider an interval system of inequalities of the form:

$$\mathbf{A}^1 \otimes x \leq \mathbf{b}^1 , \tag{6.92}$$
$$\mathbf{A}^2 \otimes x \geq \mathbf{b}^2 . \tag{6.93}$$

Theorem 6.34. *A vector $x \in \mathbb{R}^n$ is a weak solution of (6.92), (6.93) if and only if*

$$\underline{A}^1 \otimes x \leq \overline{b}^1 , \tag{6.94}$$

$$\overline{A}^2 \otimes x \geq \underline{b}^2 . \tag{6.95}$$

Proof. If (6.94), (6.95) is fulfilled, then x is a weak solution of (6.92), (6.93). Conversely, let x be a weak solution of (6.92), (6.93); i.e., there exist $A^1 \in \mathbf{A}^1$, $b^1 \in \mathbf{b}^1$, $A^2 \in \mathbf{A}^2$, $b^2 \in \mathbf{b}^2$ such that

$$A^1 \otimes x \leq b^1 ,$$
$$A^2 \otimes x \geq b^2 .$$

Then we have:

$$\underline{A}^1 \otimes x \leq A^1 \otimes x \leq b^1 \leq \overline{b}^1$$

and further

$$\overline{A}^2 \otimes x \geq A^2 \otimes x \geq b^2 \geq \underline{b}^2$$

so that x fulfills (6.94), (6.95).   □

Corollary 6.35. *A system of the form (6.92), (6.93) has a weak solution if and only if*

$$\overline{A}^2 \otimes x^*(\underline{A}^1, \overline{b}^1) \geq \underline{b}^2 , \tag{6.96}$$

where $x^*(\underline{A}^1, \overline{b}^1)$ is the principal solution of (6.94).

Proof. If (6.96) is fulfilled, then $x^*(\underline{A}^1, \overline{b}^1)$ is a weak solution of (6.92), (6.93). Then it fulfills (6.94) and thus $x \leq x^*(\underline{A}^1, \overline{b}^1)$; since x must fulfill also (6.95), we obtain:

$$\overline{A}^2 \otimes x^*(\underline{A}^1, \overline{b}^1) \geq \overline{A}^2 \otimes x \geq \underline{b}^2 ,$$

so that (6.96) is fulfilled.   □

Theorem 6.36. *A vector x is a strong solution of (6.92), (6.93) if and only if it is a solution of the system*

$$\overline{A}^1 \otimes x \leq \underline{b}^1 , \tag{6.97}$$

$$\underline{A}^2 \otimes x \geq \overline{b}^2 . \tag{6.98}$$

Proof. If x a strong solution of (6.92), (6.93), it must obviously fulfill (6.97), (6.98). Conversely, let x be a solution of (6.97), (6.98) and there exists either $A^1 \in \mathbf{A}^1$, $b^1 \in \mathbf{b}^1$ such that (6.92) is violated or $A^2 \in \mathbf{A}^2$, $b^2 \in \mathbf{b}^2$ such that (6.93) does not hold. If (6.92) is not fulfilled, there exists an index $i \in$

$\{1, \ldots, m_1\}$ such that $\sum_{j=1}^{n} {}^{\oplus} a_{ij}^1 \otimes x_j > b_i^1$ and thus $\sum_{j=1}^{n} {}^{\oplus} \overline{a}_{ij}^1 \otimes x_j > \underline{b}_i^1$ so that (6.97) is not fulfilled. If (6.93) is not satisfied, then there exists an index $i \in \{1, \ldots, m_2\}$ such that $\sum_{j=1}^{n} {}^{\oplus} a_{ij}^2 \otimes x_j < b_i^2$ and thus $\sum_{j=1}^{n} {}^{\oplus} \underline{a}_{ij}^2 \otimes x_j < \overline{b}_i^2$, so that (6.98) is not fulfilled. $\square$

The tolerance solution of (6.92), (6.93) is a vector x, which fulfills the conditions $A^1 \otimes x \in \mathbf{b}^1$, $A^2 \otimes x \in \mathbf{b}^2$, for any $A^1 \in \mathbf{A}^1$, $A^2 \in \mathbf{A}^2$. Therefore it follows from Theorem 6.27 the following.

Theorem 6.37. *A vector x is a tolerance solution of (6.92), (6.93) if and only if*

$$\underline{A}^1 \otimes x \geq \underline{b}^1 , \tag{6.99}$$

$$\overline{A}^1 \otimes x \leq \overline{b}^1 , \tag{6.100}$$

$$\underline{A}^2 \otimes x \geq \underline{b}^2 , \tag{6.101}$$

$$\overline{A}^2 \otimes x \leq \overline{b}^1 . \tag{6.102}$$

6.7 Optimization problems with $(\oplus, \otimes)$-linear interval constraints

We consider optimization problems

$$\text{minimize } f(x) = \sum_{j=1}^{n} {}^{\oplus} f_j(x_j) \tag{6.103}$$

subject to

$$A^1 \otimes x \leq \mathbf{b}^1 , \tag{6.104}$$

$$A^2 \otimes x \geq \mathbf{b}^2 , \tag{6.105}$$

$$\overline{x} \geq x \geq \underline{x}, \tag{6.106}$$

where $\mathbf{A}^1 = [\underline{A}^1, \overline{A}^1] \subset \mathbb{R}^{m_1 \times n}$, $\mathbf{A}^2 = [\underline{A}^2, \overline{A}^2] \subset \mathbb{R}^{m_2 \times n}$ and $-\infty < \underline{x}_j \leq \overline{x}_j < +\infty$ for all $j = 1, \ldots, n$, and f_j are continuous functions.

Definition 6.38. *A vector $x^{\text{opt}} \in \mathbb{R}^n$ is called the optimal weak solution of (6.103)–(6.106) if there exist $A^1 \in \mathbf{A}^1$, $A^2 \in \mathbf{A}^2$, $b^1 \in \mathbf{b}^1$, $b^2 \in \mathbf{b}^2$ such that*

$$A^1 \otimes x^{\text{opt}} \leq b^1 ,$$

$$A^2 \otimes x^{\text{opt}} \geq b^2 ,$$

$$\overline{x} \geq x^{\text{opt}} \geq \underline{x} ,$$

and $f(x^{\text{opt}}) \leq f(x)$ for any weak solution of (6.104)–(6.106).

Remark 6.39. In the other words x^{opt} from Definition 6.38 is the optimal solution among all weak solutions satisfying (6.104)-(6.106).

Taking into account Definition 6.38, Remark 6.39 and Theorem 6.34, we see that x^{opt} is in fact the optimal solution of the problem

$$\begin{aligned} &\text{minimize } f(x) \\ &\text{subject to (6.94), (6.95) and (6.106) .} \end{aligned} \tag{6.107}$$

To find it we can use the method described in Section 6.6.

Definition 6.40. *A vector x^{opt} is called the optimal tolerance solution of (6.103)-(6.106), if it is a tolerance solution of (6.104)-(6.106) and if $f(x^{\mathrm{opt}})$ $\leq f(x)$ for any tolerance solution of (6.104)-(6.106).*

Taking into account Definition 6.40, Remark 6.77 and Theorem 6.27 we see that the optimal tolerance solution of (6.103)-(6.106) is the optimal solution of the problem

$$\begin{aligned} &\text{minimize } f(x) \\ &\text{subject to (6.82), (6.83), (6.106) .} \end{aligned} \tag{6.108}$$

This solution can be again obtained by making use of the method described in Section 6.6.

Definition 6.41. *A vector x^{opt} is called the optimal strong solutionstrong solution of (6.103)-(6.106), if it is a strong solution of (6.104)-(6.106) and if $f(x^{\mathrm{opt}}) \leq f(x)$ for any strong solution of (6.104)-(6.106).*

Taking into account Definition 6.41 and Theorem 6.36, we see that the optimal strong solution of (6.103)-(6.106) can be found by solving the optimization problem

$$\begin{aligned} &\text{minimize } f(x) \\ &\text{subject to (6.86), (6.87), (6.106) .} \end{aligned} \tag{6.109}$$

We can solve this optimization problem by making use of the method from Section 6.6.

Remark 6.42. If $f_j(x_j) = c_j \otimes x_j$, where $c_j \in \mathbb{R}$, it is $f(x) = \sum_{j=1}^{n} {}^{\oplus} c_j \otimes x_j = c^T \otimes x$ and (6.103)-(6.106) can thus be called a $(\oplus, \otimes)$-linear optimization problem. Since f_js are isotone in this case, the minimization procedures from Section 6.6 can be simplified taking this fact into account (e.g., the minimum of f_j on any closed interval is attained in its left end).

Remark 6.43. It follows immediately from the preceding results that the set of weak solutions of the interval system (6.77) is equal to the set of weak solutions of the system (6.82), (6.93) if $\mathbf{A}^1 = \mathbf{A}^2 = \mathbf{A}$, $\mathbf{b}^1 = \mathbf{b}^2 = \mathbf{b}$, i.e., the system $\mathbf{A} \otimes x \leq \mathbf{b}$, $\mathbf{A} \otimes x \geq \mathbf{b}$. Therefore, we do not investigate special algorithms for optimization problems with interval $(\oplus, \otimes)$-linear equality constraints. In the sequel we investigate only conditions, which must be fulfilled by the tolerance and the strong optimal solution of an optimization problem with interval equality constraints.

Let us consider an optimization problem of the form:

$$\text{minimize } f(x) = \sum_{j=1}^{n} {}^{\oplus} f_j(x_j) \qquad (6.110)$$

subject to

$$A \otimes x = b , \qquad (6.111)$$
$$x \geq \underline{x} . \qquad (6.112)$$

In order to find the optimal tolerance solution of (6.110)–(6.112), we have to solve the following optimization problem (compare Definition 6.40 and Theorem 6.27),

$$\text{minimize } f(x) \qquad (6.113)$$

subject to

$$\underline{A} \otimes x \geq \underline{b} , \qquad (6.114)$$
$$\overline{A} \otimes x \leq \overline{b} , \qquad (6.115)$$
$$x \geq \underline{x} . \qquad (6.116)$$

This problem can be solved by making use of the algorithm described in Section 6.6. To find the optimal strong solution of (6.110)–(6.112), we have to solve—taking into account Definition 6.24 and Theorem 6.29—the optimization problem of the form

$$\text{minimize } f(x) \qquad (6.117)$$

subject to

$$\overline{A} \otimes x = \underline{b} , \qquad (6.118)$$
$$\underline{A} \otimes x = \overline{b} , \qquad (6.119)$$
$$x \geq \underline{x} . \qquad (6.120)$$

This problem can be solved again by the algorithm from Section 6.6.

Remark 6.44. If $f_j(x_j) = c_j \otimes x_j$ for all $j \in \{1, \ldots, n\}$, we obtain the $(\oplus, \otimes)$-linear optimization problem with $(\oplus, \otimes)$-linear interval equality constraints.

There arises a question of how to solve optimization problems with interval objective function of the form $\mathbf{c}^T \otimes x$, where $\mathbf{c} = [\underline{c}, \overline{c}] = \{c \mid \underline{c} \leq c \leq \overline{c}\} \subseteq \mathbb{R}^n$. Let us consider the optimization problem,

$$\text{minimize } \mathbf{c}^T \otimes x \tag{6.121}$$

$$\text{subject to } \mathbf{A}^1 \otimes x \leq \mathbf{b}^1, \ \mathbf{A}^2 \otimes x \geq \mathbf{b}^2, \ x \geq \underline{x} \ . \tag{6.122}$$

Let for any $c \in \mathbf{c}$ element $x^{\mathrm{opt}}(c)$ be the optimal weak (or tolerance or strong) solution of the problem

$$\begin{array}{c} \text{minimize } c^T \otimes x \\ \text{subject to } (6.122) \ . \end{array} \tag{6.123}$$

We interpret the problem (6.121), (6.122) as follows: Find $c^{\mathrm{opt}} \in \mathbf{c}$ such that

$$c^{\mathrm{opt}\,T} \otimes x^{\mathrm{opt}}(c^{\mathrm{opt}}) \leq c^T \otimes x^{\mathrm{opt}}(c) \text{ for any } c \in \mathbf{c}.$$

We show that $c^{\mathrm{opt}} = \underline{c}$. Since $c^T \otimes x$ is isotone both in x and in c, it holds for any fixed x and $c \in \mathbf{c}$, $\underline{c}^T \otimes x \leq c^T \otimes x$; suppose that there exists $\tilde{c} \in \mathbf{c}$ such that $\tilde{c}^T \otimes x^{\mathrm{opt}}(\tilde{c}) < \underline{c}^T \otimes x^{\mathrm{opt}}(\underline{c})$; it would be then $\underline{c}^T \otimes x^{\mathrm{opt}}(\underline{c}) \leq \underline{c}^T \otimes x^{\mathrm{opt}}(\tilde{c}) \leq \tilde{c} \otimes x^{\mathrm{opt}}(\tilde{c}) < \underline{c}^T \otimes x^{\mathrm{opt}}(\underline{c})$, which is a contradiction. Therefore $(\underline{c}, x^{\mathrm{opt}}(\underline{c}))$ is the solution of (6.123).

Remark 6.45. If $\mathbb{R} = \mathbb{R}_1$, i.e., $c_j \otimes x_j = c_j + x_j$, then the solution of (6.123) can be reduced to a problem with a noninterval objective function and interval $(\oplus, \otimes)$-linear constraints by introducing new variables $y_j = c_j + x_j$. It is then $x_j = y_j - c_j$ and $a_{ij}^{(k)} + x_j = a_{ij}^{(k)} - c_j + y_j$ for any $k = 1, 2$, $i \in \{1, \ldots, m\}$, $j \in \{1, \ldots, n\}$. Let us define $d_{ij}^{(k)} = a_{ij}^{(k)} - c_j, \underline{y}_j = c_j + \underline{x}_j$. Then it holds for $a_{ij} \in [\underline{a}_{ij}^{(k)}, \overline{a}_{ij}^{(k)}]$, $c_j \in [\underline{c}_j, \overline{c}_j]$ that $\underline{d}_{ij}^{(k)} \leq d_{ij}^{(k)} \leq \overline{d}_{ij}^{(k)}$, where $\underline{d}_{ij}^{(k)} = \underline{a}_{ij}^{(k)} - \overline{c}_j$ and $\overline{d}_{ij}^{(k)} = \overline{a}_{ij}^{(k)} - \underline{c}_j$. The original problem with the interval objective function can be therefore reduced to the problem

$$\left. \begin{array}{l} \text{minimize } \sum_{j=1}^{n} {}^{\oplus} y_j = \max_{1 \leq j \leq n} y_j \\ \text{subject to } \mathbf{D}^1 \otimes y \leq \mathbf{b}^1, \ \mathbf{D}^2 \otimes y \geq \mathbf{b}^2, \ y \geq \underline{y}_j \end{array} \right\} . \tag{6.124}$$

Problem (6.124) is a special problem of the form (6.103)–(6.106) with variables y_j and $f_j(y_j) = y_j$ for all $j = 1, \ldots, n$.

6.8 Conclusion

The results obtained in this chapter extend the possibility of applications of the $(\oplus, \otimes)$-linear systems and optimization problems to cases in which the data of the problems are inexact and the inexactness is expressed by replacing the exact coefficients of the $(\oplus, \otimes)$-linear functions involved with interval coefficients. Such problems occur, e.g., in scheduling problems with inexact processing times, reliability problems with inexact failure probabilities and others. As we have already mentioned, only problems with variables on one side of the constraints were considered here. Problems with variables on both sides of $(\oplus, \otimes)$-linear constraints may become the subject of further research.

References

1. J. Albrecht, *Monotone Iterationsfolgen und ihre Verwendung zur Lösung linearer Gleichungssysteme*, Numerische Mathematik, 3 (1961), pp. 345–358.
2. G. Alefeld and J. Herzberger, *Introduction to Interval Computations*, Academic Press, New York, 1983.
3. G. Alefeld, V. Kreinovich, and G. Mayer, *The shape of the symmetric solution set*, in Applications of Interval Computations, R. B. Kearfott and V. Kreinovich, eds., Dordrecht, 1996, Kluwer, pp. 61–79.
4. G. Alefeld, V. Kreinovich, and G. Mayer, *On the shape of the symmetric, persymmetric, and skew-symmetric solution set*, SIAM Journal on Matrix Analysis and Applications, 18 (1997), pp. 693–705.
5. G. Alefeld, V. Kreinovich, and G. Mayer, *The shape of the solution set for systems of interval linear equations with dependent coefficients*, Mathematische Nachrichten, 192 (1998), pp. 23–36.
6. G. Alefeld and G. Mayer, *On the symmetric and unsymmetric solution set of interval systems*, SIAM Journal on Matrix Analysis and Applications, 16 (1995), pp. 1223–1240.
7. P. Alexandroff and H. Hopf, *Topologie I.*, Springer Verlag, Berlin, 1935.
8. E. F. Bareiss, *Sylvester's identity and multistep integer-preserving Gaussian elimination*, Mathematics of Computation, 103 (1968), pp. 565–578.
9. H. Bauch, K.-U. Jahn, D. Oelschlägel, H. Süsse, and V. Wiebigke, *Intervallmathematik*, Teubner, Leipzig, 1987.
10. M. Baumann, *A regularity criterion for interval matrices*, in Collection of Scientific Papers Honouring Prof. Dr. K. Nickel on Occasion of his 60th Birthday, Part I, J. Garloff et al., eds., Freiburg, 1984, Albert-Ludwigs-Universität, pp. 45–50.
11. H. Beeck, *Zur Problematik der Hüllenbestimmung von Intervallgleichungssystemen*, in Interval Mathematics, K. Nickel, ed., Lecture Notes in Computer Science 29, Springer-Verlag, Berlin, 1975, pp. 150–159.
12. H. Beeck, *Linear programming with inexact data*, Technical Report TUM–ISU–7830, Technical University of Munich, Munich, 1978.
13. R. Bellman and L. Zadeh, *Decision making in fuzzy environment*, Management Science, 17 (1970), pp. 141–164.
14. F. D. Barb, A. Ben-Tal and A. Nemirovski, *Robust dissipativity of interval uncertain linear systems*, SIAM Journal on Optimization, 41 (2003), pp. 1661–1695.

15. A. Ben-Tal and A. Nemirovski, *Robust convex optimization*, Mathematics and Operations Research, 23 (1998), pp. 796–805.
16. A. Ben-Tal and A. Nemirovski, *Robust solutions of uncertain linear programs*, Operations Research Letters, 25 (1999), pp. 1–13.
17. A. Ben-Tal and A. Nemirovski, *Robust optimization - methodology and applications*, Mathematical Programming, 92 (2002), pp. 453–480.
18. A. Ben-Tal and A. Nemirovski, *On tractable approximations of uncertain linear matrix inequalities affected by interval uncertainty*, SIAM Journal on Optimization, 12 (2002), pp. 811–833.
19. A. Ben-Tal, A. Nemirovski, and C. Roos, *Robust solutions of uncertain quadratic and conic quadratic problems*, SIAM Journal on Optimization, 13 (2002), pp. 535–560.
20. A. Ben-Tal, A. Goryashko, E. Gulitzer and A. Nemirovski, *Adjustable robust solutions of uncertain linear programs*, Mathematical Programming, 99 (2004), pp. 351–376.
21. C. Bliek, *Computer Methods for Design Automation*, PhD thesis, Massachusetts Institute of Technology, Cambridge, MA, July 1992.
22. J. J. Buckley, *Possibilistic linear programming with triangular fuzzy numbers*, Fuzzy Sets and Systems, 26 (1988), pp. 135–138.
23. R. E. Burkard, B. Klinz, and R. Rudolf, *Perspectives of Monge properties in optimization*, Discrete Applied Mathematics, 70 (1996), pp. 95–161.
24. K. Cechlárová, *Solutions of interval linear systems in max-plus algebra*, in Proceedings of the 6th International Symposium on Operational Research in Slovenia, Preddvor, Slovenia, September 26-28, 2001, L. Lenart, L. Zadnik-Stirn and S. Drobne, eds., Ljubljana, 2001, Slovenian Society Informatika, Section for Operational Research, pp. 321–326.
25. K. Cechlárová and R. A. Cuninghame-Green, *Interval systems of max-separable linear equations*, Linear Algebra and Its Applications, 340/1-3 (2002), pp. 215–224.
26. S. Chanas, *Fuzzy programming in multiobjective linear programming - a parametric approach*, Fuzzy Sets and Systems, 29 (1989), pp. 303–313.
27. S. Chen and C. Hwang, *Fuzzy multiple attribute decision making*, Springer-Verlag, Berlin, Heidelberg, New York, 1992.
28. G. E. Coxson, *Computing exact bounds on elements of an inverse interval matrix is NP-hard*, Reliable Computing, 5 (1999), pp. 137–142.
29. R. A. Cuninghame-Green, *Minimax Algebra*, Lecture Notes in Economics and Mathematical Systems 166, Springer-Verlag, Berlin, 1979.
30. R. A. Cuninghame-Green and K. Cechlárová, *Residuation in fuzzy algebra and some applications*, Fuzzy Sets and Systems, 71 (1995), pp. 227–239.
31. G. Dantzig, *Linear Programming and Extensions*, Princeton University Press, Princeton, 1963.
32. A. Deif, *Sensitivity Analysis in Linear Systems*, Springer-Verlag, Berlin, 1986.
33. M. Delgado, J. Kacprzyk, J.-L. Verdegay, and M. A. Vila, *Fuzzy Optimization - Recent Advances*, Physica-Verlag, Heidelberg, New York, 1994.
34. J. Farkas, *Theorie der einfachen Ungleichungen*, Journal für die Reine und Angewandte Mathematik, 124 (1902), pp. 1–27.
35. M. Fiedler, *Special Matrices and Their Applications in Numerical Mathematics*, Martinus Nijhoff Publishers & SNTL, Dordrecht, 1986.
36. M. Fiedler, *Equilibrated anti-Monge matrices*, Linear Algebra and Its Applications, 335 (2001), pp. 151–156.

37. M. Fiedler and V. Pták, *On matrices with non-positive off-diagonal elements and positive principal minors*, Czechoslovak Mathematical Journal, 12 (1962), pp. 382–400.
38. S. Filipowski, *On the complexity of solving feasible systems of linear inequalities specified with approximate data*, Mathematical Programming, 71 (1995), pp. 259–288.
39. J. C. Fodor and M. Roubens, *Fuzzy Preference Modelling and Multi-Criteria Decision Support*, Kluwer Academic Publishers, Dordrecht-Boston-London, 1994.
40. D. Gale, H. W. Kuhn, and W. Tucker, *Linear programming and the theory of games*, in Activity Analysis of Production and Allocation, T. C. Koopmans, ed., Cowles Commission for Research in Economics, Monograph Nr. 13, John Wiley and Sons, New York, 1951, pp. 317–329.
41. M. R. Garey and D. S. Johnson, *Computers and Intractability: A Guide to the Theory of NP-Completeness*, Freeman, San Francisco, 1979.
42. J. Garloff, *Totally nonnegative interval matrices*, in Interval Mathematics 1980, K. Nickel, ed., Academic Press, New York, 1980, pp. 317–327.
43. W. Gerlach, *Zur Lösung linearer Ungleichungssysteme bei Störung der rechten Seite und der Koeffizientenmatrix*, Mathematische Operationsforschung und Statistik, Series Optimization, 12 (1981), pp. 41–43.
44. G. H. Golub and C. F. van Loan, *Matrix Computations*, The Johns Hopkins University Press, Baltimore, 1996.
45. M. Gondran and M. Minoux, *Linear algebra in dioids: A survey of recent results*, Annals of Discrete Mathematics, 19 (1984), pp. 147–164.
46. F. Gray, *Pulse code communication*. United States Patent Number 2,632,058. March 17, 1953.
47. E. Hansen, *On linear algebraic equations with interval coefficients*, in Topics in Interval Analysis, E. Hansen, ed., Oxford University Press, Oxford, 1969, pp. 33–46.
48. E. R. Hansen, *Bounding the solution of interval linear equations*, SIAM Journal on Numerical Analysis, 29 (1992), pp. 1493–1503.
49. G. Heindl, *Some inclusion results based on a generalized version of the Oettli-Prager theorem*, Zeitschrift für Angewandte Mathematik und Mechanik, Supplement 3, 76 (1996), pp. 263–266.
50. J. Herzberger and D. Bethke, *On two algorithms for bounding the inverse of an interval matrix*, Interval Computations, 1 (1991), pp. 44–53.
51. N. J. Higham, *Accuracy and Stability of Numerical Algorithms*, SIAM, Philadelphia, 1996.
52. A. J. Hoffman, *On simple linear programming problems*, in Convexity, Proceedings of Symposia in Pure Mathematics, AMS, Providence, RI, 1961, pp. 317–327.
53. R. A. Horn and C. R. Johnson, *Matrix Analysis*, Cambridge University Press, Cambridge, 1985.
54. R. A. Horn and C. R. Johnson, *Topics in Matrix Analysis*, Cambridge University Press, Cambridge, 1991.
55. A. Ingleton, *A problem in linear inequalities*, Proceedings of the London Mathematical Society, 16 (1966), pp. 519–536.
56. M. Inuiguchi, H. Ichihashi, and Y. Kume, *Some properties of extended fuzzy preference relations using modalities*, Information Sciences, 61 (1992), pp. 187–209.

57. M. Inuiguchi, H. Ichihashi, and Y. Kume, *Modality constrained programming problems: A unified approach to fuzzy mathematical programming problems in the setting of possibility theory*, Information Sciences, 67 (1993), pp. 93–126.

58. M. Inuiguchi and J. Ramík, *Possibilistic linear programming: A brief review of fuzzy mathematical programming and a comparison with stochastic programming in portfolio selection problem*, Fuzzy Sets and Systems, 111 (2000), pp. 3–28.

59. M. Inuiguchi and T. Tanino, *Scenario decomposition approach to interactive fuzzy numbers in possibilistic linear programming problems*, in Proceedings of EFDAN'99, R. Felix, ed., Dortmund, 2000, FLS Fuzzy Logic Systeme GmbH, pp. 133–142.

60. K.-U. Jahn, *Eine Theorie der Gleichungssysteme mit Intervallkoeffizienten*, Zeitschrift für Angewandte Mathematik und Mechanik, 54 (1974), pp. 405–412.

61. C. Jansson, *Zur linearen Optimierung mit unscharfen Daten*, PhD thesis, University of Kaiserslautern, 1985.

62. C. Jansson, *A self-validating method for solving linear programming problems with interval input data*, Computing Supplementum, 6 (1988), pp. 33–46.

63. C. Jansson, *On self-validating methods for optimization problems*, in Topics in Validated Computations, J. Herzberger, ed., North-Holland, Amsterdam, 1994, pp. 381–438.

64. C. Jansson, *Calculation of exact bounds for the solution set of linear interval systems*, Linear Algebra and Its Applications, 251 (1997), pp. 321–340.

65. C. Jansson and S. M. Rump, *Rigorous solution of linear programming problems with uncertain data*, ZOR-Methods and Models of Operations Research, 35 (1991), pp. 87–111.

66. L. V. Kantorovich, *Mathematical methods in the organization and planning of production*, Management Science, 6 (1960), pp. 550–559. Original Russian version appeared in 1939.

67. N. Karmarkar, *A new polynomial-time algorithm for linear programming*, Combinatorica, 4 (1984), pp. 373–395.

68. B. Kelling, *Methods of solution of linear tolerance problems with interval arithmetic*, in Computer Arithmetic and Enclosure Methods, L. Atanassova and J. Herzberger, eds., North-Holland, Amsterdam, 1992, pp. 269–277.

69. B. Kelling, *Geometrische Untersuchungen zur eingeschränkten Lösungsmenge linearer Intervallgleichungssysteme*, Zeitschrift für Angewandte Mathematik und Mechanik, 74 (1994), pp. 625–628.

70. B. Kelling and D. Oelschlägel, *Zur Lösung von linearen Toleranzproblemen*, Wissenschaftliche Zeitschrift TH Leuna-Merseburg, 33 (1991), pp. 121–131.

71. L. G. Khachiyan, *A polynomial algorithm in linear programming*, Doklady Akademii Nauk SSSR, 244 (1979), pp. 1193–1096.

72. V. Klee and G. Minty, *How good is the simplex algorithm?*, in Inequalities III, O. Shisha, ed., Academic Press, New York, 1972, pp. 159–175.

73. E. P. Klement, R. Mesiar, and E. Pap, *Triangular Norms*, Kluwer Academic Publishers Series Trends in Logic, Dordrecht-Boston-London, 2000.

74. J. Koníčková, *Optimalizační úlohy s nepřesnými daty*, PhD thesis, Charles University, Prague, 1996.

75. J. Koníčková, *Sufficient condition of basis stability of an interval linear programming problem*, Zeitschrift für Angewandte Mathematik und Mechanik, Supplement 3, 81 (2001), pp. S677–S678.

76. M. Kovacs, *Fuzzy linear programming with triangular fuzzy parameters*, in Identification, Modelling and Simulation, Paris, 1987, Proceedings of IASTED Conference, pp. 447–451.

77. M. Kovacs, *Fuzzy linear programming with centered fuzzy numbers*, in Fuzzy Optimization - Recent Advances, M. Delgado, J. Kacprzyk, J.-L. Verdegay, and M. A. Vila, eds., Physica-Verlag, Heidelberg-New York, 1994, pp. 135–147.

78. M. Kovacs and L. H. Tran, *Algebraic structure of centered M-fuzzy numbers*, Fuzzy Sets and Systems, 39 (1991), pp. 91–99.

79. R. Krawczyk, *Fehlerabschätzung bei linearer Optimierung*, in Interval Mathematics, K. Nickel, ed., Lecture Notes in Computer Science 29, Berlin, 1975, Springer-Verlag, pp. 215–222.

80. V. Kreinovich, A. Lakeyev, J. Rohn, and P. Kahl, *Computational Complexity and Feasibility of Data Processing and Interval Computations*, Kluwer Academic Publishers, Dordrecht, 1998.

81. Y. J. Lai and C. L. Hwang, *Fuzzy Mathematical Programming: Theory and Applications*, Springer-Verlag, Berlin, Heidelberg, New York, London, Paris, Tokyo, 1992.

82. Y. J. Lai and C. L. Hwang, *Multi-Objective Fuzzy Mathematical Programming: Theory and Applications*, Springer-Verlag, Berlin, Heidelberg, New York, London, Paris, Tokyo, 1993.

83. A. V. Lakeyev and S. I. Noskov, *A description of the set of solutions of a linear equation with intervally defined operator and right-hand side (in Russian)*, Doklady of the Russian Academy of Sciences, 330 (1993), pp. 430–433.

84. A. V. Lakeyev and S. I. Noskov, *On the set of solutions of a linear equation with intervally defined operator and right-hand side (in Russian)*, Siberian Mathematical Journal, 35 (1994), pp. 1074–1084.

85. B. Machost, *Numerische Behandlung des Simplexverfahrens mit intervallanalytischen Methoden*, Berichte der GMD 30, GMD, Bonn, 1970.

86. M. Marcus and H. Minc, *A Survey of Matrix Theory and Matrix Inequalities*, Allyn and Bacon, Boston, 1964.

87. G. Mayer and J. Rohn, *On the applicability of the interval Gaussian algorithm*, Reliable Computing, 4 (1998), pp. 205–222.

88. C. D. Meyer, *Matrix Analysis and Applied Linear Algebra*, SIAM, Philadelphia, 2000.

89. R. E. Moore, *Interval Analysis*, Prentice-Hall, Englewood Cliffs, NJ, 1966.

90. R. E. Moore, *Methods and Applications of Interval Analysis*, SIAM Studies in Applied Mathematics, SIAM, Philadelphia, 1979.

91. F. Mráz, *Interval linear programming problem*, Freiburger Intervall-Berichte 87/6, Albert-Ludwigs-Universität, Freiburg, 1987.

92. F. Mráz, *Solution function of an interval linear programming problem*, Report 90–03, The Technical University of Denmark, Lyngby, 1990.

93. F. Mráz, *On supremum of the solution function in linear programs with interval coefficients*, Research Report, KAM Series 93–236, Faculty of Mathematics and Physics, Charles University, Prague, 1993.

94. F. Mráz, *Úloha lineárního programování s intervalovými koeficienty*. West Bohemian University, 1993. Habilitationsschrift.

95. F. Mráz, *The exact lower bound of optimal values in interval LP*, in Scientific Computing and Validated Numerics, G. Alefeld, A. Frommer and B. Lang, eds., Mathematical Research, Vol. 90, Berlin, 1996, Akademie Verlag, pp. 214–220.

96. F. Mráz, *Calculating the exact bounds of optimal values in LP with interval coefficients*, Annals of Operations Research, 81 (1998), pp. 51–62.

97. K. G. Murty, *On the number of solutions to the complementarity problem and spanning properties of complementary cones*, Linear Algebra and Its Applications, 5 (1972), pp. 65–108.

98. J. Nedoma, *Vague matrices in linear programming*, Annals of Operations Research, 47 (1993), pp. 483–496.

99. J. Nedoma, *Sign-stable solutions of column-vague linear equation systems*, Reliable Computing, 3 (1997), pp. 173–180.

100. J. Nedoma, *Inaccurate linear equation systems with a restricted-rank error matrix*, Linear and Multilinear Algebra, 44 (1998), pp. 29–44.

101. J. Nedoma, *On solving vague systems of linear equations with pattern-shaped columns*, Linear Algebra and Its Applications, 324 (2001), pp. 107–118.

102. J. Nedoma, *Positively regular vague matrices*, Linear Algebra and Its Applications, 326 (2001), pp. 85–100.

103. A. Neumaier, *Linear interval equations*, in Interval Mathematics 1985, K. Nickel, ed., Lecture Notes in Computer Science 212, Springer-Verlag, Berlin, 1986, pp. 109–120.

104. A. Neumaier, *Tolerance analysis with interval arithmetic*, Freiburger Intervall-Berichte 86/9, Albert-Ludwigs-Universität, Freiburg, 1986.

105. A. Neumaier, *Interval Methods for Systems of Equations*, Cambridge University Press, Cambridge, 1990.

106. A. Neumaier, *A simple derivation of the Hansen-Bliek-Rohn-Ning-Kearfott enclosure for linear interval equations*, Reliable Computing, 5 (1999), pp. 131–136.

107. K. Nickel, *Die Überschätzung des Wertebereichs einer Funktion in der Intervallrechnung mit Anwendungen auf lineare Gleichungssysteme*, Computing, 18 (1977), pp. 15–36.

108. S. Ning and R. B. Kearfott, *A comparison of some methods for solving linear interval equations*, SIAM Journal on Numerical Analysis, 34 (1997), pp. 1289–1305.

109. E. Nuding, *Ein einfacher Beweis der Sätze von Oettli–Prager und J. Rohn*, Freiburger Intervall–Berichte 86/9, Albert-Ludwigs-Universität, Freiburg, 1986.

110. E. Nuding and J. Wilhelm, *Über Gleichungen und über Lösungen*, Zeitschrift für Angewandte Mathematik und Mechanik, 52 (1972), pp. T188–T190.

111. W. Oettli, *On the solution set of a linear system with inaccurate coefficients*, SIAM Journal on Numerical Analysis, 2 (1965), pp. 115–118.

112. W. Oettli and W. Prager, *Compatibility of approximate solution of linear equations with given error bounds for coefficients and right-hand sides*, Numerische Mathematik, 6 (1964), pp. 405–409.

113. S. A. Orlovsky, *Decision making with fuzzy preference relation*, Fuzzy Sets and Systems, 1 (1978), pp. 155–167.

114. S. A. Orlovsky, *On formalization of a general fuzzy mathematical programming problem*, Fuzzy Sets and Systems, 3 (1980), pp. 311–321.

115. M. Padberg, *Linear Optimization and Extensions*, Springer-Verlag, Berlin, 1999.

116. S. Poljak and J. Rohn, *Radius of nonsingularity*, Research Report, KAM Series 88–117, Faculty of Mathematics and Physics, Charles University, Prague, December 1988.

117. S. Poljak and J. Rohn, *Checking robust nonsingularity is NP-hard*, Mathematics of Control, Signals, and Systems, 6 (1993), pp. 1–9.

118. J. Ramík, *Extension principle in fuzzy optimization*, Fuzzy Sets and Systems, 19 (1986), pp. 29–37.

119. J. Ramík, *An application of fuzzy optimization to optimum allocation of production*, in Proceedings of the International. Workshop on Fuzzy Set Applications, J. Kacprzyk and S. A. Orlovsky, eds., Academia-Verlag, Laxenburg, Berlin, 1987, IIASA, pp. 227–241.

120. J. Ramík, *A unified approach to fuzzy optimization*, in Proceedings of the 2nd IFSA Congress, M. Sugeno, ed., Tokyo, 1987, IFSA, pp. 128–130.

121. J. Ramík, *Fuzzy preferences in linear programming*, in Interactive Fuzzy Optimization and Mathematical Programming, M. Fedrizzi and J. Kacprzyk, eds., Springer-Verlag, Berlin-Heidelberg-New York, 1990, pp. 114–122.

122. J. Ramík, *Linear programming with inexact and fuzzy coefficients*. Faculty of Mathematics and Physics, Charles University, 1991. Habilitationsschrift.

123. J. Ramík, *Vaguely interrelated coefficients in lp as bicriterial optimization problem*, International Journal on General Systems, 20 (1991), pp. 93–114.

124. J. Ramík, *Inequality relations between fuzzy data*, in Modelling Uncertain Data, H. Bandemer, ed., Akademie Verlag, Berlin, 1992, pp. 158–162.

125. J. Ramík, *Some problems of linear programming with fuzzy coefficients*, in Operation Research Proceedings: Papers of the 21st Annual Meeting of DGOR 1992, K.-W. Hansmann, A. Bachem, M. Jarke, and A. Marusev, eds., Springer-Verlag, Heidelberg, 1993, pp. 296–305.

126. J. Ramík, *New interpretation of the inequality relations in fuzzy goal grogramming problems*, Central European Journal for Operations Research and Economics, 4 (1996), pp. 112–125.

127. J. Ramík, *Linear programming with inexact coefficients: Some results in computational complexity*, Research Report, Japan Advanced Institute for Science and Technology, Hokuriku, 1999.

128. J. Ramík, *Fuzzy goals and fuzzy alternatives in fuzzy goal programming problems*, Fuzzy Sets and Systems, 111 (2000), pp. 81–86.

129. J. Ramík and K. Nakamura, *Canonical fuzzy numbers of dimension two*, Fuzzy Sets and Systems, 54 (1993), pp. 167–180.

130. J. Ramík, K. Nakamura, I. Rozenberg, and I. Miyakawa, *Joint canonical fuzzy numbers*, Fuzzy Sets and Systems, 53 (1993), pp. 29–47.

131. J. Ramík and H. Rommelfanger, *A single- and multi-valued order on fuzzy numbers and its use in linear programming with fuzzy coefficients*, Fuzzy Sets and Systems, 57 (1993), pp. 203–208.

132. J. Ramík and H. Rommelfanger, *Fuzzy mathematical programming based on some new inequality relations*, Fuzzy Sets and Systems, 81 (1996), pp. 77–88.

133. J. Ramík and H. Rommelfanger, *A new algorithm for solving multi-objective fuzzy linear programming problems*, Foundations of Computing and Decision Sciences, 3 (1996), pp. 145–157.

134. J. Ramík and M. Vlach, *Generalized concavity as a basis for optimization and decision analysis*, Tech. Report IS-RR-2001-003, Japan Advanced Institute for Science and Technology, Hokuriku, March 2001.

135. J. Ramík and M. Vlach, *Generalized concavity in optimization and decision making*, Kluwer Academic Publishers, Boston-Dordrecht-London, 2001.

136. J. Ramík and J. Římánek, *Inequality relation between fuzzy numbers and its use in fuzzy optimization*, Fuzzy Sets and Systems, 16 (1985), pp. 123–138.

137. J. Ramík and J. Římánek, *The linear programming problem with vaguely formulated relations between the coefficients*, in Interfaces between Artificial Intelligence and Operations Research in Fuzzy Environment, M. Fedrizzi, J. Kacprzyk, and S. A. Orlovsky, eds., D.Riedel, Dordrecht-Boston-Lancaster-Tokyo, 1989, pp. 104–119.

138. H. Ratschek and W. Sauer, *Linear interval equations*, Computing, 28 (1982), pp. 105–115.

139. J. Renegar, *Some perturbation theory for linear programming*, Mathematical Programming, 65 (1994), pp. 73–91.

140. G. Rex, *Zum Regularitätsnachweis von Matrizen*, Zeitschrift für Angewandte Mathematik und Mechanik, 75 (1995), pp. S549–S550.

141. G. Rex and J. Rohn, *A note on checking regularity of interval matrices*, Linear and Multilinear Algebra, 39 (1995), pp. 259–262.

142. G. Rex and J. Rohn, *Sufficient conditions for regularity and singularity of interval matrices*, SIAM Journal on Matrix Analysis and Applications, 20 (1999), pp. 437–445.

143. F. N. Ris, *Interval Analysis and Applications to Linear Algebra*, PhD thesis, Oxford University, Oxford, 1972.

144. J. Rohn, *A perturbation theorem for linear equations*. To appear.

145. J. Rohn, *Soustavy lineárních rovnic s intervalově zadanými koeficienty*, Ekonomicko-matematický obzor, 12 (1976), pp. 311–315.

146. J. Rohn, *Input-output planning with inexact data*, Freiburger Intervall-Berichte 78/9, Albert-Ludwigs-Universität, Freiburg, 1978.

147. J. Rohn, *Duality in interval linear programming*, in Interval Mathematics 1980, K. Nickel, ed., Academic Press, New York, 1980, pp. 521–529.

148. J. Rohn, *Interval linear systems with prescribed column sums*, Linear Algebra and Its Applications, 39 (1981), pp. 143–148.

149. J. Rohn, *Strong solvability of interval linear programming problems*, Computing, 26 (1981), pp. 79–82.

150. J. Rohn, *Interval linear systems*, Freiburger Intervall-Berichte 84/7, Albert-Ludwigs-Universität, Freiburg, 1984.

151. J. Rohn, *Proofs to "Solving interval linear systems"*, Freiburger Intervall-Berichte 84/7, Albert-Ludwigs-Universität, Freiburg, 1984.

152. J. Rohn, *Solving interval linear systems*, Freiburger Intervall-Berichte 84/7, Albert-Ludwigs-Universität, Freiburg, 1984.

153. J. Rohn, *Miscellaneous results on linear interval systems*, Freiburger Intervall-Berichte 85/9, Albert-Ludwigs-Universität, Freiburg, 1985.

154. J. Rohn, *Inner solutions of linear interval systems*, in Interval Mathematics 1985, K. Nickel, ed., Lecture Notes in Computer Science 212, Springer-Verlag, Berlin, 1986, pp. 157–158.

155. J. Rohn, *On sensitivity of the optimal value of a linear program*, Ekonomicko-matematický obzor, 25 (1989), pp. 105–107.

156. J. Rohn, *Systems of linear interval equations*, Linear Algebra and Its Applications, 126 (1989), pp. 39–78.

157. J. Rohn, *Characterization of a linear program in standard form by a family of linear programs with inequality constraints*, Ekonomicko-matematický obzor, 26 (1990), pp. 71–73.

158. J. Rohn, *Interval solutions of linear interval equations*, Aplikace matematiky, 35 (1990), pp. 220–224.

159. J. Rohn, *An existence theorem for systems of linear equations*, Linear and Multilinear Algebra, 29 (1991), pp. 141–144.

160. J. Rohn, *Cheap and tight bounds: The recent result by E. Hansen can be made more efficient*, Interval Computations, 4 (1993), pp. 13–21.

161. J. Rohn, *Stability of the optimal basis of a linear program under uncertainty*, Operations Research Letters, 13 (1993), pp. 9–12.

162. J. Rohn, *NP-hardness results for some linear and quadratic problems*, Technical Report 619, Institute of Computer Science, Academy of Sciences of the Czech Republic, Prague, January 1995.

163. J. Rohn, *Complexity of some linear problems with interval data*, Reliable Computing, 3 (1997), pp. 315–323.

164. J. Rohn, *Linear programming with inexact data is NP-hard*, Zeitschrift für Angewandte Mathematik und Mechanik, Supplement 3, 78 (1998), pp. S1051–S1052.

165. J. Rohn, *Computing the norm $\|A\|_{\infty,1}$ is NP-hard*, Linear and Multilinear Algebra, 47 (2000), pp. 195–204.

166. J. Rohn, *Solvability of systems of linear interval equations*, SIAM Journal on Matrix Analysis and Applications, 25 (2003), pp. 237–245.

167. J. Rohn and V. Kreinovich, *Computing exact componentwise bounds on solutions of linear systems with interval data is NP-hard*, SIAM Journal on Matrix Analysis and Applications, 16 (1995), pp. 415–420.

168. J. Rohn and J. Kreslová, *Linear interval inequalities*, Linear and Multilinear Algebra, 38 (1994), pp. 79–82.

169. H. Rommelfanger, *Entscheiden bei Unschärfe - Fuzzy Decision Support Systeme*, Springer-Verlag, Berlin-Heidelberg, 1988.

170. H. Rommelfanger and R. Slowinski, *Fuzzy linear programming with single or multiple objective functions*, in Fuzzy Sets in Decision Analysis, Operations Research and Statistics, R. Slowinski, ed., Kluwer Academic, Boston-Dordrecht-London, 1998, pp. 179–213.

171. S. M. Rump, *Solving algebraic problems with high accuracy*, in A New Approach to Scientific Computation, U. Kulisch and W. Miranker, eds., Academic Press, New York, 1983, pp. 51–120.

172. S. M. Rump, *On the solution of interval linear systems*, Computing, 47 (1992), pp. 337–353.

173. S. M. Rump, *Verification methods for dense and sparse systems of equations*, in Topics in Validated Computations, J. Herzberger, ed., North-Holland, Amsterdam, 1994, pp. 63–135.

174. M. Sakawa and H. Yano, *Interactive decision making for multiobjective programming problems with fuzzy parameters*, in Stochastic Versus Fuzzy Approaches to Multiobjective Mathematical Programming Under Uncertainty, R. Slowinski and J. Teghem, eds., Kluwer Academic, Dordrecht, 1990, pp. 191–220.

175. H. Samelson, R. Thrall, and O. Wesler, *A partition theorem for Euclidean n-space*, Proceedings of the American Mathematical Society, 9 (1958), pp. 805–807.

176. S. P. Shary, *O nekotorykh metodakh resheniya lineinoi zadachi o dopuskakh*, Preprint 6, Siberian Branch of the Soviet Academy of Sciences, Krasnoyarsk, 1989.

177. S. P. Shary, *A new class of algorithms for optimal solution of interval linear systems*, Interval Computations, 2 (1992), pp. 18–29.

178. S. P. Shary, *On controlled solution set of interval algebraic systems*, Interval Computations, 6 (1992), pp. 66–75.
179. S. P. Shary, *Solving the tolerance problem for interval linear systems*, Interval Computations, 2 (1994), pp. 6–26.
180. S. P. Shary, *Solving the linear interval tolerance problem*, Mathematics and Computers in Simulation, 39 (1995), pp. 53–85.
181. S. P. Shary, *Algebraic approach to the interval linear static identification, tolerance and control problems, or One more application of Kaucher arithmetic*, Reliable Computing, 2 (1996), pp. 3–33.
182. S. P. Shary, *Algebraic solutions to interval linear equations and their applications*, in Numerical Methods and Error Bounds, G. Alefeld and J. Herzberger, eds., Mathematical Research, Vol. 89, Akademie Verlag, Berlin, 1996, pp. 224–233.
183. S. P. Shary, *Controllable solutions sets to interval static systems*, Applied Mathematics and Computation, 86 (1997), pp. 185–196.
184. S. P. Shary, *A new technique in systems analysis under interval uncertainty and ambiguity*, Reliable Computing, 8 (2002), pp. 321–418.
185. V. V. Shaydurov and S. P. Shary, *Resheniye interval'noi algebraicheskoi zadachi o dopuskakh*, Preprint 5, Siberian Branch of the Soviet Academy of Sciences, Krasnoyarsk, 1988.
186. Y. I. Shokin, *Interval'nyj analiz*, Nauka, Novosibirsk, 1981.
187. Y. I. Shokin, *On interval problems, interval algorithms and their computational complexity*, in Scientific Computing and Validated Numerics, G. Alefeld, A. Frommer, and B. Lang, eds., Akademie Verlag, Berlin, 1996, pp. 314–328.
188. A. L. Soyster, *Convex programming with set-inclusive constraints, and application to inexact LP*, Operations Research, 21 (1973), pp. 155–157.
189. A. L. Soyster, *A duality theory for convex programming with set-inclusive constraints*, Operations Research, 22 (1974), pp. 892–898.
190. A. A. Vatolin, *On linear programming problems with interval coefficients (in Russian)*, Zhurnal Vychislitel'noi Matematiki i Matematicheskoi Fiziki, 24 (1984), pp. 1629–1637.
191. J. R. Vera, *Ill-posedness and the complexity of deciding existence of solutions to linear programs*, SIAM Journal on Optimization, 6 (1996), pp. 549–569.
192. J. von Neumann, *On a Maximization Problem*, Institute of Advanced Studies, Princeton, 1947. Manuscript.
193. N. N. Vorob'ev, *Extremal algebra of positive matrices (in Russian)*, Elektronische Informationsverarbeitung und Kybernetik, 3 (1967), pp. 39–70.
194. E. W. Weisstein, *Gray code*. http://mathworld.wolfram.com/GrayCode.html.
195. B. Werners, *Interactive fuzzy programming system*, Fuzzy Sets and Systems, 23 (1987), pp. 131–147.
196. R. R. Yager, *On a general class of fuzzy connectives*, Fuzzy Sets and Systems, 4 (1980), pp. 235–242.
197. H.-J. Zimmermann, *Fuzzy programming and linear programming with several objective functions*, Fuzzy Sets and Systems, 1 (1978), pp. 45–55.
198. K. Zimmermann, *Extremal algebra (in Czech)*, Research Report EML 46, Institute of Economics, Czechoslovak Academy of Sciences, Prague, 1976.
199. K. Zimmermann, *On max-separable optimization problems*, Annals of Discrete Mathematics, 19 (1984), pp. 357–362.
200. K. Zimmermann, *Disjunctive optimization, max-separable problems and extremal algebras*, Theoretical Computer Science, 293 (2003), pp. 45–54.

List of Symbols

$\mathbf{A}$	set of matrices; in particular, an interval matrix
$|A|$	absolute value of a matrix (componentwise)
$\underline{A}$	lower bound of an interval matrix $\mathbf{A} = [\underline{A}, \overline{A}]$
$\overline{A}$	upper bound of an interval matrix $\mathbf{A} = [\underline{A}, \overline{A}]$
$[A]_\alpha$	α-cut of fuzzy set A
$(A)_\alpha$	strict α-cut of fuzzy set A
A_c	midpoint matrix of an interval matrix $\mathbf{A} = [A_c - \Delta, A_c + \Delta]$
$\alpha \circ a$	scalar multiplication
$\alpha \wedge \beta$	$= \min\{\alpha, \beta\}$
$\tilde{a}^{\mathrm{L}}(\alpha), \tilde{a}^{\mathrm{R}}(\alpha)$	left and right end-point of α-cut of fuzzy quantity $\tilde{a}$
$A_{i.}$	ith row of A
$A_{.j}$	jth column of A
A^{-1}	inverse matrix
$A(\mathcal{M}, \mathcal{N})$	submatrix
A^+	the Moore–Penrose inverse of A
$[A/A_{11}]$	Schur complement
A^T	transpose of A
A_{yz}	$= A_c - T_y \Delta T_z$
$A \le B$	$A_{ij} \le B_{ij}$ for each i, j
$A < B$	$A_{ij} < B_{ij}$ for each i, j
$\mathbf{b}$	set of vectors; in particular, an interval vector
$\underline{b}$	lower bound of an interval vector $\mathbf{b} = [\underline{b}, \overline{b}]$
$\overline{b}$	upper bound of an interval vector $\mathbf{b} = [\underline{b}, \overline{b}]$
b_c	midpoint vector of an interval vector $\mathbf{b} = [b_c - \delta, b_c + \delta]$
b_y	$= b_c + T_y \Delta$
$\mathbb{C}$	the set of complex numbers
$\mathbb{C}^{m \times n}$	$m \times n$ complex matrices
$\mathbb{C}^n$	complex vector space
card	number of elements
$\mathrm{Cl}(X)$	closure of X

codet	codeterminant
Conv X	the convex hull of X
Core(A)	the core of fuzzy set A
$\mathcal{C}(S)$	the complement of the set S
δ	radius vector of an interval vector $\mathbf{b} = [b_c - \delta, b_c + \delta]$
Δ	radius matrix of an interval matrix $\mathbf{A} = [A_c - \Delta, A_c + \Delta]$
det	determinant
e	$= (1, 1, \ldots, 1)^T$
e_j	jth column of the unit matrix I
$f(A, b, c)$	optimal value of a linear programming problem
$\underline{f}(\mathbf{A}, \mathbf{b}, \mathbf{c})$	lower bound of the range of the optimal value of an interval linear programming problem
$\overline{f}(\mathbf{A}, \mathbf{b}, \mathbf{c})$	upper bound of the range of the optimal value of an interval linear programming problem
$\mathcal{F}_I(\mathbb{R})$	set of all fuzzy intervals of $\mathbb{R}$
$\mathcal{F}_N(\mathbb{R})$	set of all fuzzy numbers of $\mathbb{R}$
$\mathcal{F}_0(\mathbb{R})$	set of all fuzzy quantities of $\mathbb{R}$
$\mathcal{F}(X)$	set of all fuzzy subsets of X
g_2	2-norm
$\mathcal{H}$	Hankel matrix
Hgt(A)	height of fuzzy set A
I	unit (or identity) matrix
$\underset{\sim}{\leq}^{\min}$	Min-fuzzy extension of relation $\leq$
$\underset{\sim}{\leq}^{\max}$	Max-fuzzy extension of relation $\leq$
$<, >$	bilinear form
$\mathcal{L}, \mathcal{R}$	generating functions of $(\mathcal{L}, \mathcal{R})$-fuzzy interval
$\mathcal{L}^{(-1)}, \mathcal{R}^{(-1)}$	pseudo-inverse functions of $\mathcal{L}$ and $\mathcal{R}$
MaxU, MinU	sets of maximal and minimal elements of $\mathbf{U}$ in $\mathbb{R}^m$
$\max\{A, B\}$	componentwise maximum of matrices (vectors)
$\min\{A, B\}$	componentwise minimum of matrices (vectors)
$\mu_A : \mathbb{R}^m \to [0, 1]$	membership function of fuzzy subset A of $\mathbb{R}^m$
$\oplus, \otimes$	extremal algebra operations
$\|x\|$	length, norm
(x, y)	inner product
$\overline{\varphi}(\mathbf{A}, \mathbf{b}, \mathbf{c})$	optimal value of an auxiliary problem (p. 92)
$\mathbb{R}$	the set of real numbers
$\mathbb{R}^{m \times n}$	$m \times n$ real matrices
$\mathbb{R}^n$	real vector space
$\mathbb{R}^n_+$	nonnegative cone (orthant) in $\mathbb{R}^n$
$r(A)$	rank of A
$\varrho(A)$	spectral radius of A
sgn x	sign vector of a vector x
$\sigma(P)$	sign of the permutation P

$\sim$	similar
$\mathrm{Supp}(A)$	the support of fuzzy set A
$\mathcal{T}$	Toeplitz matrix
T_L, S_L	Lukasiewicz t-norm, bounded sum t-conorm
T_M, S_M	Minimum t-norm, Maximum t-conorm
$\mathrm{tr}(A)$	trace of A
T_y	diagonal matrix with diagonal vector y
x_y	unique solution of the equation $A_c x - T_y \Delta |x| = b_y$
Y_m	the set of all ± 1-vectors in $\mathbb{R}^m$

Index